Urban Street Stormwater Guide

About Island Press

Since 1984, the nonprofit Island Press
has been stimulating, shaping, and
communicating the ideas that are
essential for solving environmental
problems worldwide. With more than
800 titles in print and some 40 new
releases each year, we are the nation's
leading publisher on environmental
issues. We identify innovative
thinkers and emerging trends in the
environmental field. We work with
world-renowned experts and authors to
develop cross-disciplinary solutions to
environmental challenges. Island Press
designs and implements coordinated
book publication campaigns in order
to communicate our critical messages
in print, in person, and online using the
latest technologies, programs, and
the media. Our goal: to reach targeted
audiences—scientists, policymakers,
environmental advocates, the media,
and concerned citizens—that can and
will take action to protect the plants
and animals that enrich our world, the
ecosystems we need to survive, the
water we drink, and the air we breathe.
Island Press gratefully acknowledges the
support of its work by the Agua Fund,
Inc., The Margaret A. Cargill Foundation,
Betsy and Jesse Fink Foundation, The
William and Flora Hewlett Foundation,
The Kresge Foundation, The Forrest
and Frances Lattner Foundation, The
Andrew W. Mellon Foundation, The
Curtis and Edith Munson Foundation,
The Overbrook Foundation, The
David and Lucile Packard Foundation,
The Summit Foundation, Trust for
Architectural Easements, The Winslow
Foundation, and other generous donors.
The opinions expressed in this book
are those of the authors and do not
necessarily reflect the views of our
donors.

Urban
Street
Stormwater
Guide

National Association of
City Transportation Officials

**National Association of
City Transportation Officials**

ABOUT NACTO

NACTO's mission is to build cities as places for people, with safe, sustainable, accessible and equitable transportation choices that support a strong economy and vibrant quality of life.

The National Association of City Transportation Officials is a 501(c)(3) nonprofit association that represents large cities on transportation issues of local, regional, and national significance. The organization facilitates the exchange of transportation ideas, insights, and best practices among large cities, while fostering a cooperative approach to key issues facing cities and metropolitan areas. As a coalition of city transportation departments, NACTO is committed to raising the state of practice for street design and transportation by building a common vision, sharing data, peer-to-peer exchange in workshops and conferences, and regular communication among member cities.

**National Association of
City Transportation Officials**
120 Park Avenue, 23rd Floor
New York, NY 10017
www.nacto.org

© Copyright 2017 National Association of City Transportation Officials

ISBN: 978-1-61091-812-1

Library of Congress Catalog Control Number: 2017930448

NACTO EXECUTIVE BOARD

Seleta Reynolds, President
General Manager, Los Angeles
Department of Transportation

Scott Kubly, Vice President
Director, Seattle Department
of Transportation

Crissy Fanganello, Secretary
Director of Transportation for Public Works,
City and County of Denver

Danny Pleasant, Treasurer
Director of Transportation, City of Charlotte

Joseph Barr, Affiliate Member Representative
Director, Traffic, Parking & Transportation, City
of Cambridge

Janette Sadik-Khan, Chair
Principal, Bloomberg Associates

NACTO MEMBER CITIES

Atlanta
Austin
Baltimore
Boston
Charlotte
Chicago
Denver
Detroit
Houston
Los Angeles
Minneapolis
New York
Philadelphia
Phoenix
Pittsburgh
Portland
San Antonio
San Diego
San Francisco
San José
Seattle
Washington, DC

**TRANSIT AFFILIATE
MEMBERS**
King County Metro
Los Angeles Metro
Miami-Dade County
New York MTA
Portland TriMet

**AFFILIATE
MEMBERS**
Arlington, VA
Boulder, CO
Burlington, CT
Cambridge, MA
Chattanooga, TN
El Paso, TX
Fort Lauderdale, FL
Hoboken, NJ
Indianapolis, IN
Louisville, KY
Madison, WI
Memphis, TN
Miami Beach, FL
Nashville, TN
New Haven, CT
Oakland, CA
Palo Alto, CA
Salt Lake City, UT
San Luis Obispo, CA
Santa Monica, CA
Somerville, MA
Vancouver, WA
Ventura, CA
West Hollywood, CA

**INTERNATIONAL
MEMBERS**
Montréal, QC
Toronto, ON
Vancouver, BC

NACTO PROJECT TEAM

Linda Bailey
Executive Director

Corinne Kisner
Director of Policy and Special Projects

Matthew Roe
Director, Designing Cities Initiative

Aaron Villere
Program Associate,
Designing Cities Initiative

Alex Engel
Senior Program Associate

Craig Toocheck
Program Analyst/Designer,
Designing Cities Initiative

CONTRIBUTORS

Shanti Colwell, Seattle Public Utilities

Susan McLaughlin, Seattle Department of
Transportation

Lacy Shelby, Minneapolis Department
of Community Planning and Economic
Development

Ariel Ben-Amos, Philadelphia Water
Department

Peg Staeheli, MIG | SvR

Kathy Gwilym, MIG | SvR

Amalia Leighton, MIG | SvR

TECHNICAL REVIEW

Technical review conducted by American Society of Civil Engineers' Environmental & Water Resources Institute: ASCE's technical source for environmental and water-related issues. 1801 Alexander Bell Drive, Reston, VA 20191. www.asce.org

ACKNOWLEDGMENTS

The *Urban Street Stormwater Guide* would not have been possible without the support of the Summit Foundation and Seattle Public Utilities. Many thanks especially to Darryl Young of the Summit Foundation and Shanti Colwell of Seattle Public Utilities. The project team owes tremendous thanks to all of the members of the steering committee, as well as contributors and representatives of partner organizations, who generously gave their time, expertise, and energy to the development of this guide. Thanks especially to Scott Struck and Brian Parsons of ASCE's EWRI for their partnership and expertise.

Special thanks to Heather Boyer and the staff at Island Press for bringing this publication into print.

Contents

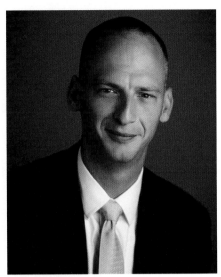

Scott Kubly
NACTO Vice President
Director, Seattle Department of
Transportation

Mami Hara
General Manager/CEO, Seattle Public Utilities

Water is a resource, not a problem.

For too long, cities have treated water as waste. Stormwater has been viewed as something to be "managed," and streets have prevented stormwater from doing exactly what we most need it to do: get back into the ground. But just like building new highways doesn't solve traffic congestion, adding capacity to traditional 'gray' stormwater systems is proving to be financially, and ecologically, untenable. And with a changing climate bringing more frequent and intense storms, heat waves, and droughts, pressure is mounting to protect against flooding risks, improve water quality, and build resilient urban places.

The urgency of this challenge requires us to make better use of our resources. Both transportation mobility and stormwater management are rooted in physical street space. A siloed approach results in competing priorities and conflicting project timelines. The streets and sidewalks that disrupt natural hydrology with an impermeable surface of asphalt and concrete must be reimagined as sites for sustainable stormwater infrastructure. Instead of competing for street space, our transportation and water management agencies must work together to find value in stormwater and integrate ecology into urban life.

Here in Seattle, we understand that complete streets that incorporate natural drainage solutions make our city a better, more resilient place. Investments in green stormwater infrastructure that support a healthy water cycle are paired with geometric changes in the right-of-way that calm traffic and make room for walking, biking, and public space. A partnership between the Seattle Department of Transportation (SDOT) and Seattle Public Utilities (SPU) allows us to capitalize more fully on opportunities in the public right-of-way. Green Streets are codified in the Land Use Code and require private developments to incorporate green stormwater infrastructure into their design. Complete Streets, mandated by ordinance in 2007, requires SDOT to assess each capital project's opportunity to incorporate green stormwater infrastructure. These two regulatory tools result in projects that create functional green spaces that contribute to a more vibrant streetscape.

We're not alone in putting stormwater to better use. Cities around the country are institutionalizing partnerships, creating holistic programs, and adopting best practices in stormwater street design. From Philadelphia's Green City Clean Waters program, to Denver's *Ultra-Urban Green Infrastructure Guidelines*, to Portland's Grey to Green initiative, we're seeing intentional approaches to sustainable stormwater management centered on the role of streets.

The *Urban Street Stormwater Guide* synthesizes this wealth of local experience into a national resource for sustainable design. The result of a nationwide collaboration between city transportation and stormwater experts, the guide illustrates how streets of every size can incorporate stormwater management techniques alongside sustainable, multi-modal mobility.

From bioretention swales in transit boarding islands, to stormwater curb extensions that shorten crossing distances for pedestrians, to protected bike lanes built with permeable pavement, we can, and must, put our public assets to work, holding our infrastructure to a higher standard. The *Urban Street Stormwater Guide* provides the tools we need to do so. Together, we can find value in water and design better cities.

Scott Kubly
NACTO Vice President
Director, Seattle Department of
Transportation

Mami Hara
General Manager/CEO, Seattle Public
Utilities

About the Guide

The *Urban Street Stormwater Guide* is a resource to reintroduce valuable ecological processes into the life of urban streets. As more and more cities reclaim street space for human life and habitat, and enact far-reaching plans and strategies to address climate change, this guide presents a new playbook for utilizing existing and new infrastructure to manage rainfall and reconnect ecosystems. Developed by a new network of NACTO city water, transportation, and public works departments, the *Urban Street Stormwater Guide* provides practitioners, leaders, advocates, and stakeholders with the tools to reintegrate water into the street.

Using the Guide

PURPOSE & ORIGIN

The *Urban Street Stormwater Guide* provides planning and design guidance for green stormwater infrastructure on city streets, and for the design and engineering of stormwater management practices that support goals for improving mobility and streets as human places while reducing the impacts of stormwater runoff and human activity on natural ecological processes. The guide has been developed on the basis of other design guidance, as well as city case studies, best practices in urban environments, research and evaluation of existing designs, and professional consensus. These sources, as well as the specific designs and elements included in the guide, are based on North American street design practice. The content of the guide has been developed collaboratively by practitioners and professionals across the United States.

STRUCTURE

The contents of the *Urban Street Stormwater Guide* are presented in a non-linear fashion, suitable for reference during the design process. Internal cross-references, a list of further resources by topic, and endnotes are provided to assist the reader in developing a deep understanding of the subject.

The Stormwater Streets chapter assembles elements presented in greater depth throughout the guide, with "Stormwater Elements" sections providing the greatest level of detail. Some renderings illustrate potential subsurface conditions and designs.

Some sections of the guide include a **CONTEXT** or **APPLICATION** discussion. The specific applications are provided for reference and include common existing uses, rather than an exhaustive or exclusive lists of all potential uses.

For most topics and treatments in this guide, the reader will find three levels of guidance:

» **CRITICAL** features are elements for which there is a strong consensus of absolute necessity.

» **RECOMMENDED** features are elements for which there is a strong consensus of added value. Most dimensions and other parameters that may vary, as well as accommodations that are desirable but not universally feasible, are included in this section to provide some degree of flexibility.

» **OPTIONAL** features are elements that may vary across cities and may add value, depending on the situation.

Note: Certain sections contain only a general **DISCUSSION** section and have no **CRITICAL**, **RECOMMENDED**, or **OPTIONAL** points.

Dimension guidance is sometimes presented in multiple levels within the guide, to be applied based on the specific needs and constraints of real streets on a case-by-case basis.

» **MINIMUM DIMENSIONS** are presented for use in geometrically constrained conditions. Bioretention cells or other elements that use minimum dimensions will typically not provide a high-performing facility. Nonetheless, minimum dimensions often allow stormwater facilities to be constructed where space constraints and competing uses are present, especially when retrofitting existing streets.

» **DESIRED MINIMUM DIMENSIONS** provide basic functional space in normal conditions. Larger dimensions are generally encouraged and can have performance or maintainability benefits. In other respects, desired minimum dimensions are similar to the lower end of recommended dimensions.

» **RECOMMENDED DIMENSIONS** provide for comfortable siting and performance in many common conditions. Where a range of dimensions is provided, choose a dimension based on location, context, and local experience. In some cases, such as corner radii, larger than recommended dimensions are less safe. However, if presented with factors not considered in the guidance, smaller or larger dimensions may perform better than recommended dimensions.

» **MAXIMUM DIMENSIONS** typically refer to motor vehicle traffic facilities. Exceeding maximum dimensions may result in safety issues and should be carefully weighed in context.

GUIDE CONTEXT

This guide focuses on green infrastructure design within urban streets and rights-of-way. The guide does not address green stormwater management strategies on private property, such as roofs and parking lots, nor does it address drainage and infiltration around controlled-access highways.

This guide aims to bridge the gaps between city staff working in transportation, public works, and water departments. It seeks to balance multi-modal mobility and environmental performance. Street design guidance assumes that users of all modes are present, and that city streets must provide accommodation for multi-modal users regardless of age or ability. Specific conditions such as on-street parking or loading, some driveways, and a moderate-to-high volume of movement on foot or on bicycle are assumed in most contexts. Sidewalks and pedestrian crossings are assumed to exist in some form in all cases.

For complementary information on designing safe streets for walking, bicycling, and taking transit, readers are referred to other existing NACTO publications, including:

> *Urban Street Design Guide*
> *Urban Bikeway Design Guide*
> *Transit Street Design Guide*
> *Global Street Design Guide*

The treatments and topics discussed in this guide must be tailored to individual situations and contexts. NACTO encourages good engineering judgment in all cases. Decisions should be thoroughly documented. To assist with this, this guide links to references and cites relevant materials and studies.

RELATION TO OTHER GUIDANCE

Several major national guidance documents exist that are relevant to green infrastructure and street design in ways that overlap with the NACTO *Urban Street Stormwater Guide.*

Specific standards in the ***Americans with Disabilities Act Accessibility Guidelines*** (ADAAG)—developed by the US Access Board, adopted by the US Department of Justice and Department of Transportation—are cited where applicable.

The US Access Board's **Public Rights-Of-Way Accessibility Guidelines** (PROWAG)—proposed in 2011, under consideration for adoption as the US standard as of publication, and recommended as a best practice guide by FHWA—include detailed accessibility guidance developed specifically for streets. PROWAG differs in some cases from the ADAAG, generally calling for wider facilities than ADAAG, and is cited where dimensions or approaches differ.

The US Environmental Protection Agency enforces the Clean Water Act, including the **National Pollutant Discharge Elimination System**, and authorizes states and local jurisdictions to create integrated plans for managing stormwater and reducing pollutants in watersheds. This guide discusses strategies to meet local or regional NPDES requirements to reduce combined sewer overflows and sanitary flows.

Many cities have developed local guidance for both complete streets design and green stormwater infrastructure. Local guidance documents may include sizing and modeling criteria, as well as complete streets policies and green infrastructure standard designs and manuals. NACTO references materials from a selection of these guides and urges municipalities to use the *Urban Street Stormwater Guide* as a basis for creating or updating local standards.

While this guide discusses permeable pavements as a street design tool for stormwater management, ASCE's **Permeable Pavements Recommended Design Guidelines** provides detailed technical guidance and information for choosing, sizing, and engineering using permeable paving materials. This guide refers to that document while providing additional discussion on mobility and implementation issues in urban contexts.

The Manual on Uniform Traffic Control Devices (MUTCD) does not regulate stormwater facilities, but, along with state and local standards, may specify whether a particular sign, signal, or marking should be used in a given design context. Users can check the status of specific traffic control elements included in this guide on the FHWA website.

Geometric design features that separate motor vehicle traffic from other uses of the roadway, including curbs and vertical and horizontal elements, are not traffic control devices and are not regulated by the MUTCD.

Fairmount Avenue & N 3rd Street, **PHILADELPHIA, PA**

1   Streets as Ecosystems

Green Street Principles

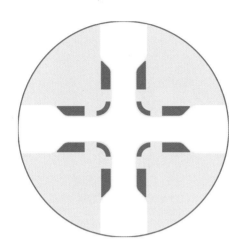

PROTECT AND RESTORE NATURAL RESOURCES

Capturing, filtering, and infiltrating stormwater is critical in urban environments where impervious surface covers 60% or more of all land area. Sustainable stormwater infrastructure filters pollutants from water and restores the natural hydrological cycle, protecting water resources.[1] Green infrastructure also improves air quality, mitigates the urban heat island effect, and increases species habitat, from small oases for birds and insects to the large water bodies that eventually receive stormwater runoff.

PROMOTE HEALTH, EQUITY, & HUMAN HABITAT

Green streets are part of healthy, equitable urban design that views streets as vital public spaces. Incorporating green elements into streets improves mental and physical health through better air quality, valuable shade, beautification, and contact with nature in areas where access to parks is limited. Ensure that the benefits of green stormwater infrastructure are provided equitably, especially in neighborhoods that have historically borne disproportionate air and water pollution or that lack green space.

DESIGN FOR SAFETY & MOBILITY

Street reconstruction projects that incorporate green infrastructure should be aligned with citywide traffic safety and mobility efforts, especially where opportunities arise to move curbs and reallocate street space for people walking and biking. Green infrastructure can be leveraged in conjunction with other street design projects to realize complementary goals, including transit access and safe mobility, providing greater value from city projects.

DESIGN FOR LIFE CYCLE

Green stormwater infrastructure is an asset for cities, providing quantifiable financial benefits. Stormwater management strategies should be planned and implemented with consideration for life-cycle costs and benefits, including the potential impacts of climate change and storm events. Green street elements that are properly designed, operated, and maintained extend the useful life of other infrastructure, especially graywater systems and pavement surfaces.

DESIGN FOR RESILIENCE

As the intensity and frequency of storms increases in many cities, and as drought conditions intensify in other cities, sustainable stormwater management is critical for climate change mitigation and adaptation. Incorporating natural systems into the built environment promotes ecosystem health and urban resilience.

OPTIMIZE FOR PERFORMANCE

Green stormwater infrastructure should be implemented at a network scale, but must be tailored to the specifics of its site. Use an understanding of topography and microclimates, available space, accessibility needs and the many human functions of a street, and desirable infiltration capacity to design appropriate green stormwater systems. Use the street to restore connections to the natural water cycle, and make comprehensive, citywide investments to see watershed-level benefits.

Vine Street, **SEATTLE, WA**

Streets are Ecosystems

The opportunity is ripe to reimagine how streets function in cities, not just as mobility corridors and public spaces, but as part of the natural ecosystem. With re-urbanization, aging infrastructure, and a changing climate, sustainable stormwater management is a core challenge for resilient cities.

Historically, streets have formed an impermeable paved layer on top of green space, disrupting hydrological cycles and requiring expensive stormwater infrastructure to manage stormwater runoff and protect ground and surface water quality. As cities face storm events of increasing frequency, duration, and intensity, as well as more persistent drought conditions, it is time to ask more of our streets.

Urban streets can reconnect rainfall to the environmental life of the city. Forward-thinking planners, engineers, and designers are treating streets as part of the ecological fabric of cities, integrating green infrastructure into the street alongside transit infrastructure and safe places for people walking and biking. By thinking of streets as ecosystems, we can build cities that are more resilient, sustainable, and enjoyable places to live.

Why Sustainable Stormwater Management Matters

Cities are defined by water. Waterways define city edges and boundaries, shape growth and development, and provide essential resources for human populations and the built environment. However, development patterns have too often removed water from urban places, channeling stormwater out of the human environment and therefore restricting natural functions and ecosystem services at great economic expense.

In the past, stormwater has been treated as waste, and stormwater management has meant dispensing of runoff as quickly as possible after a rainfall. This approach has required expensive "graywater" infrastructure: concrete and metal pipes, gutters, tanks, and treatment plants to convey, detain, and treat stormwater before discharging it into local water bodies. In many cities, intense storms overwhelm the gray infrastructure system, resulting in an outfall of polluted water into nearby streams and rivers, and potentially causing impassable streets and flooded homes and businesses. In cities across the country, gray infrastructure systems are under-maintained and reaching the end of their useful lives. Replacing this aging infrastructure can be a prohibitively expensive proposition.[2]

Such a singular approach to stormwater management is no longer possible or desirable. In an age of climate change, urbanization, and increasingly frequent, and intense storms and prolonged, devastating droughts, cities are now treating stormwater as a resource to be valued, not waste to be managed.

SEATTLE, WA

Green stormwater infrastructure (GSI) reintroduces ecological functions back into the built environment. Soil-water-plant systems—including biofiltration planters, bioretention swales, trees, and permeable pavements—intercept stormwater before it reaches gray infrastructure. Some water is infiltrated into the ground, some is evaporated into the air, and some is temporarily stored before being slowly released into the sewer system. Green stormwater infrastructure helps to reduce runoff volume to gray infrastructure and filter pollutants, protecting water quality and mitigating risks of flooding. Investments in green stormwater infrastructure complement gray infrastructure and may extend the useful life of major capital street and sewer projects. In addition to its hydrological role, green stormwater infrastructure can offer valuable co-benefits, like calming traffic and beautifying the urban landscape. An integrated approach to green stormwater management in the public right-of-way is central to the design of resilient urban landscapes.

The High Cost of Conventional Infrastructure

In 2010, New York City estimated that updating the city's stormwater system to control combined sewer overflows using only gray infrastructure would cost the city **$6.8 billion** of capital investment over twenty years. By blending gray and green strategies, the city reduced its estimated cost by $1.5 billion.[3]

Economic Losses from Storms

In 2016, four flooding events and eight severe storms have each incurred damages exceeding **$1 billion** across the United States.[4]

Public Health Risks

860 municipalities across the US with a total population of 40 million people, have combined sewer systems.[5] During heavy rainfall events that overburden the gray infrastructure system, untreated stormwater and sewage flows directly into local water bodies, causing serious public health risks and environmental pollution.

Prevalence of Urban Flooding

Storms of all magnitudes can cause flooding of homes and businesses, disrupting lives and damaging property. A study of Cook County, Illinois, found that urban flooding is chronic and systemic; property owners suffered an **average of $6,000 in damages per flooding event**, and 87% of homeowners surveyed had experienced multiple flooding events.[6] In addition to damaging buildings, flooding can disrupt street operations and prevent safe transportation.

Frequency of Storm Surges

Climate change is causing temperature increases and sea level rise, which combine to create increasingly frequent and dramatic storm surges that threaten low-lying parts of the city. In many coastal cities, "once-in-a-century" storm surges may occur once a decade in the future.[7]

The Role of Streets

Cities are uniquely positioned to take action on sustainable stormwater management.

Concrete and asphalt dominate urban landscapes. Typically in urbanized areas, 60% of land or more is impervious surface.[8] Water that falls on roofs, streets, and parking lots cannot soak into the ground, and instead becomes stormwater runoff, collecting pollutants like oil, grease, heavy metals, and bacteria before flowing through gutters and storm drains, and eventually discharging into local water bodies.

Streets comprise one-third or more of all land and half of the impervious surface in many cities.

Streets are the interstitial spaces that enable cities; they provide a network for all of the dynamic social, economic, and physical activities that make cities vital human habitat. By design, streets channel and convey stormwater, providing a network along which all the rain that falls on the city can be routed. While streets have traditionally functioned to collect and drain stormwater to water treatment facilities and designated outfalls, streets that capture and infiltrate stormwater back into the urban ecosystem can generate enormous ecological, economic, and public health benefits.

Streets present both a barrier to natural hydrology and an enormous opportunity for a better approach to stormwater management. Public rights-of-way are controlled by city agencies, from design to construction to operations to regular maintenance and permitting. Interdepartmental coordination enables more streamlined and holistic projects, ensuring that streets not only collect and infiltrate stormwater, but also realize the potential health, safety, and mobility benefits of urban stormwater street design. Integrated design strategies address water quality and regulatory compliance along with traffic calming, bike and pedestrian access, safety, urban greening and aesthetic improvements, air quality, urban temperature, public health, community development, and equity.

Streets can be changed; the time to act is now.

Existing Condition

Auto-oriented streets have large linear swaths of underutilized and impermeable space. **NEW YORK, NY**

Interim Redesign

Interim projects to test or rapidly implement geometric redesigns can introduce temporary or movable green features, like planters, that improve the human environment. Interim projects may be implemented with a local maintenance partner. **NEW YORK, NY**

Capital Reconstruction

Interim designs can be formalized during full capital reconstruction, with improvements extended to drainage, public space, street furniture, and permanent vegetation and tree plantings. Transportation, Water & Sewer, and Parks departments may all have formal roles throughout design and implementation. **NEW YORK, NY**

Complete Streets are Green Streets

A flooded street is not a complete street. During storm events, people walking, bicycling, and using transit are the first users to encounter barriers and lose access to the street, and are the last to regain it. Green street design tools, which integrate stormwater control and management within the right-of-way, are a critical component of complete street design, ensuring the street remains usable and safe for all people during storm events, regardless of mode.

Take into consideration both the impacts of stormwater on multi-modal travel and the potential for green street investments to transform the public realm and create economic, social, and environmental benefits for all street users.

Street User	User Considerations	GSI Benefits & Solutions
1 People Walking	» Ponding of stormwater, especially near intersection crossings and ramps, creates barriers, especially for people using mobility devices. Ponding may result from blocked drains and basins, wear over time to roadway slope and pavement quality, or improperly designed stormwater drainage systems. For people using mobility devices, stormwater on the street functionally and significantly prevents access. » Large or fast runoff streams also create barriers and degrade walking comfort. » Drainage grates, lips, high storm drains, and large seams sited in or near pedestrian crossings introduce hazards.	» Greenery and trees—especially those that introduce shade canopy—make the walking environment more inviting and pleasant by reducing temperature, attenuating noise, and improving air quality. » Green infrastructure can be used to calm traffic and improve safety conditions. » High-quality public gathering spaces with natural features improve mental health, and create opprtunities for community development and social cohesion.
2 People Using Transit	» People riding transit are also pedestrians and interact similarly with stormwater. Puddles or streams can impede walking and wheelchair access to transit stations and bus stops. » Rider comfort is enhanced by shelter, shade, and greenscape at the transit stop. Improving rider comfort and experience is critical to growing transit as a mode.	» GSI can be integrated into transit facilities, including boarding bulbs and islands, to improve passenger comfort and natural drainage near stops. » Transit shelter and facility roofs—usually owned or overseen by public agencies—can incorporate green features.
3 People Bicycling	» Puddling or ponding of stormwater impedes safe and enjoyable bicycling where drainage is insufficient or ineffective. » Wet pavement may discourage some potential riders who are concerned about mud and spray. An extended drying or drainage period may displace bicycle trips into other transportation modes. » The details of stormwater infrastructure design are safety-critical: poorly placed or antiquated drainage grates and storm drains can pose hazards to people biking, including slick surfaces, debris around grates, and the potential for wheels to become stuck in grates.	» Green stormwater infrastructure can be incorporated alongside bikeways to improve drainage and increase bicycling comfort and access during and after storms of any size. » Permeable pavements can be implemented on bikeways and raised cycle tracks to reduce the period of time required for pavement to dry. » Planters or vegetation may be incorporated into protected bikeway buffer elements to increase rider comfort and reduce stress.
4 People Driving Motor Vehicles	» Flooded streets can become impassable for motor vehicles. Puddles and pooled water can create poor or dangerous driving conditions, with splashing, poor visibility due to reflections, and unpredictable swerving to avoid water. » Poorly draining streets hinder curbside access for vehicle entry and loading.	» Green infrastructure facilities that capture runoff and reduce flooding and ponding promote safer driving conditions. » Design and site green infrastructure with sensitivity to context, and implement GSI with other geometric changes that reduce vehicle speed and improve visibility. People driving cars, especially in adverse weather, at night, or when driving at an unsafe speed, may drive their vehicle into a stormwater facility. Incursions that damage stormwater infrastructure are costly to repair.

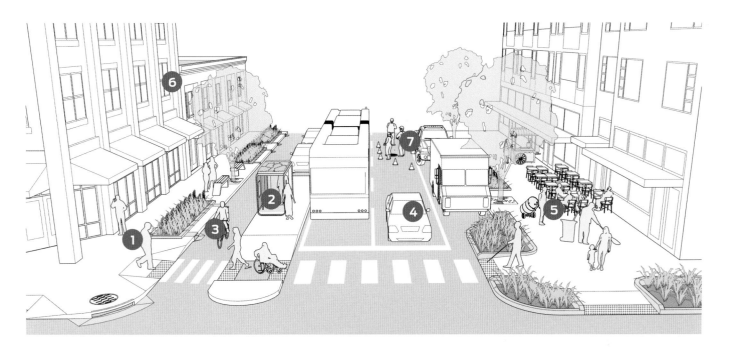

Street User	User Considerations	GSI Benefits & Solutions
5 People Conducting Business	» Curbside access is universally critical, regardless of travel mode or trip purpose; people making freight deliveries or doing business by foot, bike, handtruck, transit, or motor vehicle all need to access the curb in order to reach their destinations. » Freight movement and deliveries are essential to businesses and cities' economies, requiring thoughtful integration into street design and urban life. Flooded streets that impede freight movement take an economic toll.	» The success and vitality of commercial districts and neighborhood storefronts depend upon the ability of workers, visitors, and essential services to be able to access and use streets comfortably. » Economic performance is tied to the comfort and attractiveness of streets—urban environments with green expressions, from planters to street trees to stormwater infrastructure, perform better than streets without green improvements.
6 People Residing	» Insufficient stormwater management on streets can cause flooding in homes and businesses. Property owners incur financial losses from flooded buildings, and insurance rates can rise after repeated claims. » Chronically wet houses and basements can reduce property values and deter potential buyers. Frequent flooding can cause mold, which can lead to an increase in respiratory problems » People may use downstream water bodies for recreational activities. Poor water quality in lakes, rivers, and streams poses a public health risk and limits opportunities to use waterfronts for recreation.	» The presence of green stormwater infrastructure can be an asset to property owners. Green stormwater networks work with gray infrastructure to mitigate flood risk, especially with careful siting guidelines and design strategies near basements and subsurface structures.[9] » Street trees and greenscape have been shown to increase property values.[10] » Green infrastructure can be implemented in collaboration with private properties to direct right-of-way runoff to bioretention areas beyond the right-of-way. » Runoff from buildings and structures can be captured and infiltrated into right-of-way green infrastructure.
7 People Working / Performing Maintenance	» City crews and utility companies require periodic access to elements within the street to perform routine or emergency maintenance, such as sewers, cleanouts, and subsurface utility lines. » Pavements cuts impact drainage and accessibility. » Snow clearance and storage during winter months impact street operations.	» Green infrastructure must be designed with maintenance in mind; crews must be able to access and navigate equipment around green elements. » Green infrastructure must be implemented with consideration for existing or planned subsurface lines (see page 24). » Vegetated strips provide linear space for snow storage.

Lawrence Street, **CHICAGO, IL**

2 Planning for Stormwater

21st Street, **PASO ROBLES, CA**

Developing a Sustainable Stormwater Network

Sustainable stormwater management works to reconcile natural hydrology with human land use and development. Typically, a city's stormwater management strategy is driven by federal regulatory requirements, existing sewer infrastructure, and the regional climate and ecology. Plans and strategies take shape within the context of the street, where transportation and utility infrastructure compete for space and zoning codes shape development patterns.

Integrating green stormwater infrastructure into the right-of-way requires a coordinated approach and a holistic vision for sustainable urban design. Planning a stormwater network concurrently with an active transportation network unlocks new opportunities for cities and their streets.

Setting Goals for Stormwater Management

A city's stormwater management goals are generally rooted in regulatory requirements for water quality under the Clean Water Act. City objectives and strategies vary based on the type and capacity of the sewer system, the risk of local flooding, and the need to comply with National Pollutant Discharge Elimination System permits.

Combined Sewer System

Some cities discharge their stormwater into pipes that also receive and convey wastewater or sanitary sewer flows. These types of systems are called combined sewer systems and are most commonly found in older cities; new developments do not build combined sewer systems.

Combined sewer systems are typically connected to wastewater treatment plants that then discharge treated water into a receiving water body. However, during heavy rains, water flows can overwhelm the infrastructure capacity. When pipes and treatment plants are unable to manage the flows, water is discharged directly into receiving water bodies without being treated, in an event called a combined sewer overflow (CSO).

Depending upon an agency's permit requirements with EPA, the number of CSO events that can occur at each overflow point varies. Many agencies that have CSOs exceeding EPA's overflow thresholds are utilizing bioretention facilities and other green stormwater infrastructure (with and without gray storage facilities) to come into regulatory compliance by reducing the volume of stormwater runoff that reaches the combined system, and therefore reducing the frequency and volume of CSO events.

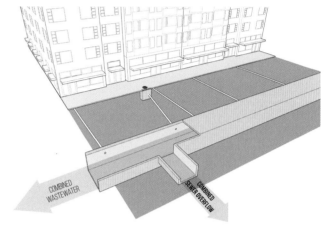

Combined sewer system

Municipal Separate Storm Sewer System (MS4)

Current gray stormwater infrastructure guidelines require that stormwater runoff be collected and conveyed separately from sanitary sewer lines. In municipal separate storm sewer systems (MS4s), stormwater runoff is often discharged into receiving water bodies with limited or no water quality treatment.

Most cities have MS4 permits, which require new development and redevelopment projects to install water quality treatment facilities and/or flow control facilities prior to discharging runoff into receiving water bodies. The permits may also require retrofits to existing sites or streets to treat and reduce runoff.

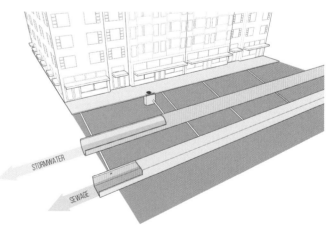

Municipal Separate Storm Sewer System

Green stormwater infrastructure is a cost-effective way to come into compliance with regulatory requirements and create other ecological and social benefits. The EPA strongly encourages the use of green infrastructure to manage stormwater and meet federal water quality requirements.[1] Green infrastructure projects are generally designed to complement gray infrastructure systems performing a combination of volume management, water quality improvement, and flood control.

VOLUME MANAGEMENT

Since the volume of stormwater runoff increases as impervious surface area increases, the two major strategies related to volume management are to increase pervious area or to divert runoff into the green infrastructure system. Green stormwater infrastructure systems are designed to convert surface area from impervious to permeable, and reduce the volume of runoff that reaches the sewer system or downstream water bodies, reducing the burden on gray infrastructure systems and infiltrating stormwater runoff directly back into the soil. Green infrastructure projects and programs set specific stormwater volume management goals, such as the first inch of rainfall on a given area.

WATER QUALITY IMPROVEMENT

Stormwater runoff from streets carries sediment, debris, chemicals, and pollutants (such as heavy metals from brake pads, oil dripping from engines, and grit from tires). Green stormwater infrastructure projects are designed to capture pollutants in runoff and prevent them from reaching downstream water bodies. Water quality treatment requirements vary based on the type of receiving water body (ocean, salt water bay, river, stream, wetland, or lake) as well as its existing condition. Green infrastructure projects often set a specific water quality goal, such as removing 80% of total suspended solids (TSS).

PEAK FLOW REDUCTION

Especially heavy rainstorms can cause combined sewer overflow events as well as flooded streets, parking lots, private property, and basements. Green stormwater infrastructure programs aim to reduce peak flow rates and mitigate flooding. Cities may design to accommodate the high runoff flows and flood risks of a given peak storm event, such as a ten-year storm (a storm that has a 10% chance of occurring in any given year).

Regional Climate & Ecology

Sustainable stormwater management aims to reconnect the natural water cycle. Green infrastructure intercepts stormwater runoff at its source and does one or more of the following, depending on project goals:

DETENTION: Collect and hold runoff in temporary storage facilities or vegetated systems before slowly releasing the water into the downstream system.

RETENTION: Capture and hold stormwater on-site to reduce runoff to sewer systems. Water is then evaporated, transpired, or infiltrated through the soil.

(BIO)FILTRATION: Remove particulate matter and other pollutants by filtering stormwater runoff through porous media such as sand, soil, or other filter.

INFILTRATION: Absorb stormwater through the ground surface and into the soil.

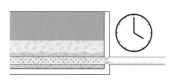  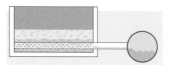 

Green infrastructure systems are sensitive to regional weather, climate, and ecology. These overarching conditions inform stormwater management needs and goals at the programmatic and project levels. Designers must consider the regional climate conditions, today and in the future.

At a most basic level, it is important to consider the amount of precipitation that falls, and the rate at which water evaporates from the surface or infiltrates back into the ground. Local weather and climate determine precipitation patterns and temperature, and soil composition affects how quickly water is absorbed into the ground. These factors determine regional sizing and modeling coefficients for runoff and infiltration.

EVAPOTRANSPIRATION RATE

Evapotranspiration rate, or the amount of water that can be evaporated or transpired from a surface over a specified duration, varies by latitude, topography, altitude, wind conditions, and time of year, as well as surface characteristics such as permeability, albedo, and types of vegetation.

In some climates, evapotranspiration varies widely through the seasons; green infrastructure should be designed with consideration for high and low expected conditions. Irrigation may be necessary during the very dry season, while fast infiltration is desirable during the very rainy seasons. In cold climates, snow storage is critical during winter.

INFILTRATION RATE

Green infrastructure elements and systems may be designed to infiltrate a specified amount of stormwater; for example, a system goal may be to infiltrate the first inch of precipitation of each storm event in a tributary area, or to match pre-development runoff conditions.

Native underlying soils affect the infiltration rate. Regions with naturally fast-draining soils, especially gravel or sandy soils, may have dramatic infiltration capacity, but require accounting for water seepage around built structures. Potential effects on local drinking water supplies may also need to be considered in areas with high infiltration capacity. In areas with shallow bedrock or clay, infiltration rates may be naturally low. A high groundwater table may also affect a city's ability to implement green stormwater infrastructure.

PLANT HEALTH

Plants and trees absorb stormwater runoff and stabilize the soil against erosion. GSI projects should use native, non-invasive, drought-tolerant vegetation. Select plants that can tolerate the inundation of stormwater runoff but also be adaptable to local climate conditions such as the dry season or severe cold conditions. In snowy climates, select plants that can tolerate salt.

Low Temperature / Low Precipitation

900 East, **SALT LAKE CITY, UT**

Recharging groundwater is necessary all year; snow pack often feeds groundwater and surface waterways through the rest of the year.

Snow clearance and storage are considerations, as is the space for operating snow plows and removal equipment.

Frequent freeze-thaw cycles make materials selection an important consideration—concrete, pavement, and markings may wear or fade more quickly.

Low Temperature / High Precipitation

University Avenue, **MINNEAPOLIS, MN**

In some regions, green infrastructure must be designed to withstand heavy snowfall. Consider winter conditions when selecting construction materials and plantings.

Snow and ice clearance is a major factor; consider the space required for operating snow plows or snow removal equipment, as well as the space required to store snow while safely maintaining mobility.

In cities with cold temperatures and frequent ice events, consider how the use of road salts to de-ice roadways will affect water quality and plant health in stormwater infrastructure.

High Temperature / Low Precipitation

North 9th Avenue, **TUSCON, AZ**

Cities in the North American West and Southwest are arid throughout the year, and may be fighting persistent drought conditions. Water conservation and reuse is a major consideration. Any stormwater that can be captured and returned to groundwater must be to maintain healthy vegetation.

Plants that can tolerate very little water are incorporated into street vegetation for their ability to withstand dry seasons. However, vegetation should be chosen to withstand flooding conditions, too.

High Temperature / High Precipitation

St Charles Avenue, **NEW ORLEANS, LA**

Cities located along coastlines or riverfronts, especially in the North American South and East Coast, typically have very rainy seasons where flooding risk increases.

These cities may rely upon surrounding wetland or deltas to attenuate flooding from nearby water bodies.

Green infrastructure systems are typically focused on reducing runoff flow and subsequent erosion and contamination that negatively impact fish and wildlife habitat.

Aligning with Goals for Streets

Streets are the physical spaces where city goals take shape. To make the most of streets, use an integrated approach to stormwater management, mobility, and land use that is both intentional and opportunistic. Coordinate street infrastructure projects and align goals to maximize the value of public investments, avoid conflicting construction schedules, and unlock complementary benefits.

Overlay stormwater network plans with plans for bicycle infrastructure, pedestrian safety improvements, transit routes or stations, and public space investments to better predict and plan for opportunities for coinvestment. Land use and zoning changes that accompany site redevelopment should include stormwater management strategies in the right-of-way alongside opportunities to improve multi-modal mobility.

Consistent and thoughtful partnerships between city agencies and other stakeholders, such as transit agencies or business improvement districts, are critical to aligning goals and developing holistic street design projects (for further information, refer to Collaboration & Partnerships on page 122).

STORMWATER NETWORK

Evaluate watershed health, existing gray sewer infrastructure, flood zones, and regulatory requirements for volume reduction or water quality to inform a citywide sustainable stormwater management plan that includes green infrastructure in the right-of-way.[2]

A network of green infrastructure can be leveraged for additional sustainability goals, including increasing park density, reducing the urban heat island effect, and connecting people with greenery.

ACTIVE TRANSPORTATION NETWORK

Automobile-oriented transportation policy results in wide streets and high impervious surface cover. Geometric changes to the street that improve conditions for people walking, bicycling, and riding transit also unlock space for green infrastructure.

Analyze motor vehicle volume and speed at a network level to identify streets that are "over-designed" for vehicles. Reallocate excess travel space recaptured by street redesigns to increase pervious surface and improve safety and comfort for all travel modes.

Biking: Implementation of high-quality bicycle facilities can be aligned with green infrastructure, including pervious pavements and bioretention facilities in buffer spaces.

Walking: Integrate GSI on sidewalks and at intersections to widen walkways, create bulbouts for intersections and potentially midblock crossings, and calm traffic by providing visual cues to motorists to reduce travel speeds, creating safer and more pleasant conditions for pedestrians.

Transit: Major investments in high-capacity transit such as light rail can open connected linear space for GSI. At a smaller scale, spot treatments for better transit service, like boarding bulbs and islands, can include greenscape and improve the rider experience. Coordination with the transit operator is critical for mutual success.

LAND USE & IMPERVIOUS SURFACE

Zoning codes and land use regulations dictate development patterns and impervious surface cover. Streets, roofs, and parking lots generate runoff if stormwater is not managed on-site.

Zoning Codes: Consider the role of zoning codes in creating impervious surface cover and influencing travel behavior, such as by requiring a minimum number of parking spaces for residential or commercial buildings.

Regulations and Incentives: Enact or revise land use regulations to encourage or require the use of green infrastructure to manage stormwater runoff on-site. Use development incentives to address stormwater that falls on building roofs and private property.

Revise zoning codes to encourage walking, biking, and transit and reduce impervious surface cover.

Sustainable Development: Connect green infrastructure in the right-of-way with off-site infiltration galleries, such as public parks and plazas where there is more available space. Consider how specific land uses affect stormwater strategies. For example, current or redeveloped industrial land uses may require environmental remediation to remove pollutants from the soil and watershed prior to implementation of green infrastructure.

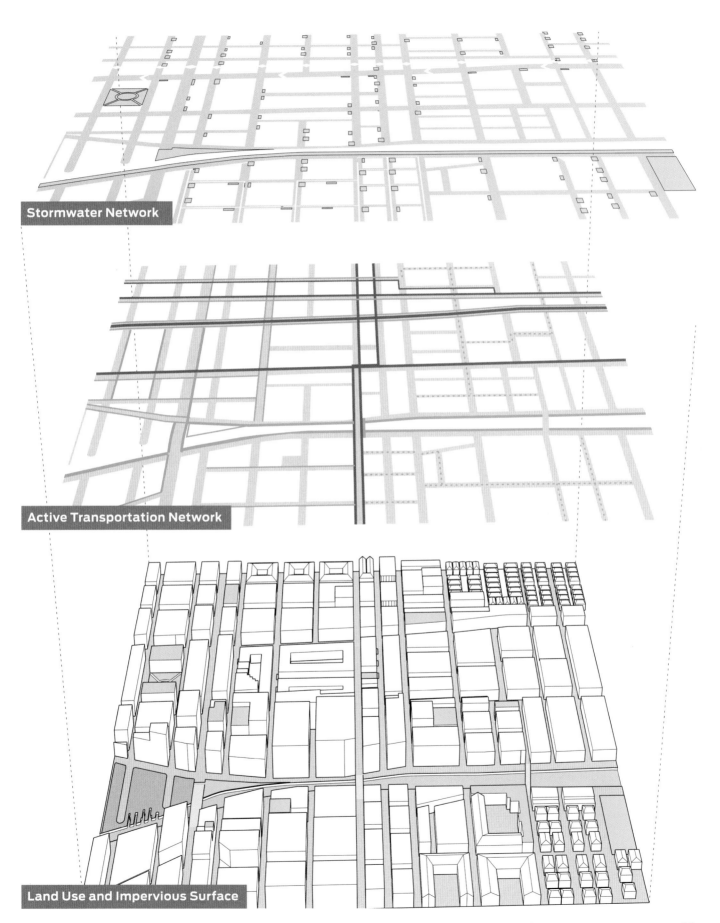

Stormwater Network

Active Transportation Network

Land Use and Impervious Surface

Case Study: Greenways to Rivers Arterial Stormwater System (GRASS)

Location: Los Angeles, CA

Context: Citywide Planning Tool

Participating Agencies: Los Angeles Bureau of Sanitation, Watershed Protection Division; Department of Landscape Architecture, California State Polytechnic University, Pomona; Department of Landscape Architecture, University of California, Los Angeles Extension; USC Price School of Public Policy

Timeline: *GRASS Phase I:* 2013
GRASS Phase II: 2017

GOALS

Capture stormwater: Capture and infiltrate runoff produced by 0.75 inches of rainfall (the design storm in LA, the 85th percentile event) and decrease the peak runoff rate during extreme rain events.

Reduce park poverty: Connect park-poor areas of the city to an active greenway network with ecological, social, and economic benefits.

Provide planning goals: Identify priorities in building out the greenway network, so that future projects will provide the greatest stormwater capture potential, connectivity, and access to park-poor communities.

Potential greenway connections between urban landmarks and the Los Angeles River, **LOS ANGELES, CA**[3]

OVERVIEW

The City of Los Angeles's Bureau of Sanitation worked with three landscape architecture studios at nearby universities to develop a GIS methodology and tool for addressing LA's urgent water needs, including reliance on remote water sources, persistent drought, and pollution of the Los Angeles River and Pacific Ocean.

The resulting tool, GRASS (Greenways to Rivers Arterial Stormwater System), provides a citywide green infrastructure solution aimed at managing, treating, reusing, and recharging groundwater throughout the region, while providing recreational space in the form of greenways to disadvantaged communities.

ISSUES

The majority (approximately 88%) of Los Angeles's water is imported from sources hundreds of miles away, by an aging infrastructure system serving under extended drought conditions.

At the same time, 75% of Los Angeles's alluvial soils are covered with impermeable surfaces, preventing significant groundwater recharge during rain events and circumventing natural treatment processes.

Precipitation in developed areas becomes urban runoff, which conveys various sources of pollution. LA's stormwater flood control system sends this runoff and pollutants directly into the Los Angeles River, which feeds into the Pacific Ocean via local beaches and harbors. Unmitigated, this scenario wastes an invaluable resource and poses human health and economic risks.

At the same time, much of Los Angeles suffers from a lack of parkland and open spaces, with fewer than one-third of LA's children living within walking distance of a park or playground, revealing an urgent need for more equitable distribution of green spaces.

DESIGN DETAILS

The GRASS Team looked at existing street classifications with wide corridors, bike routes, bus routes, and existing storm drains to create a Regional Greenway Network, which also includes utility and tributary channel easements.

Local agencies and practitioners examined major destinations (schools, parks, and civic institutions), and helped determine priority areas of the Greenway Network. Priority corridors that provide major connections to popular destination points across the grid were identified and used to model the storage capacity of the system.

In addition to modeling the network, arterials were subclassified into five groups to indicate the cross-sectional storage capacities of greenways that could be built based on various soil infiltration rates. The GRASS I greenways were analyzed in terms of their potential to alleviate park poverty, filter industrial stormwater, capture and use or infiltrate water, or serve as a connector between gaps in the network. The GRASS II priority greenways modeled the stormwater capture and treatment potential of fully connective and restored tributaries within the network, and aid project developers in planning for capturing the first 0.75" of runoff from the public right-of-way in a stormwater event.

At the watershed scale, the GRASS approach to building connectivity between projects will assist individual project developers to balance uses (such as irrigation demands) with capture and storage (as may be required for LID mitigation).

While project-scale designs will vary depending on the soils classification of each part of the network, common details include curbside bioretention swales to capture runoff, percolating through various levels of biofiltration before reuse or reaching an underground storage system. Stored water in this system can be used to irrigate trees, supporting tree canopy in a region lacking in it. In extreme storm events, excess runoff in these "offline" systems and infiltration chambers add capacity to the existing stormwater system, and provide increased flood safety, reduced peak flows, and reduced pollutants downstream.

KEYS TO SUCCESS

Use funding opportunities to accomplish multiple goals.
GRASS is an ambitious tool that proposes restoration of natural stream benefits through integrated projects and connectivity, allowing for transformational opportunities for neighborhoods across the city. Increasing access to recreational opportunities and open space while accomplishing stormwater infiltration goals also supports regulatory compliance.

Visible public benefits are crucial. In
order to gain community and political support, visible benefits are needed for the first phase of ambitious projects. This helps provide momentum for future projects.

Prioritize corridors. With a citywide
plan, it is not possible to tackle every potential project at once. A holistic plan that clearly outlines priority corridors in a way that provides connectivity will provide the most benefits in an efficient timeframe.

OUTCOMES

Developed a mechanism and tool for restoring natural tributary functions through connectivity.
The tool identified priority areas and classifications that will enable Los Angeles to build a greenway system that is interconnected, equitable, and high-performing.

Developed a subwatershed masterplan. To illustrate applications, a "high priority" subwatershed, Hazeltine, and four high-priority regional corridors were identified as high-impact case studies for developing the potential of the regional plan.

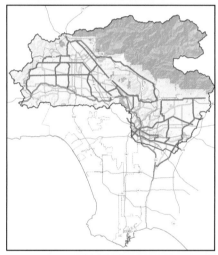

GRASS I Greenways Network

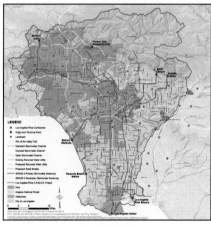

GRASS II Primary and Seconday Greenways Network

Solving the Street Design Puzzle

After identifying network-level goals and opportunities for stormwater, mobility, and land use, evaluate specific street contexts to guide project-level decisions.

At each potential site, consider native soil quality, existing infrastructure, adjacent buildings and structures, slope, and citywide policy goals.

Choose from a range of stormwater elements and techniques depending on the space available—from bioretention planters to stormwater trees. Use a combination of stormwater elements to meet overall project goals.

Design tools and elements are further detailed in Chapter 4 Stormwater Elements, on page 75.

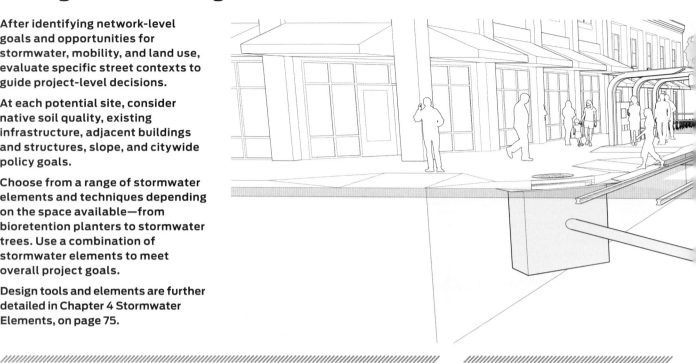

NATIVE SOIL & GROUNDWATER

Native soil characteristics determine the types of facilities that are possible, as well as the required intensity of the design. Capitalize on high-quality soils for their infiltration capacity.

Analyze the soil's infiltration rate. Gravelly and sandy soils infiltrate water more quickly than clay. In areas with slow infiltration or poor soil quality, GSI can be designed with an underdrain.

Test soil for contamination from previous projects, especially at industrial sites.

Investigate the depth to the groundwater table. In areas with a shallow groundwater table or seasonal high water table, use non-infiltration stormwater facilities to prevent pollutants from reaching the water table.

EXISTING INFRASTRUCTURE

Consider the location of existing gray stormwater infrastructure, especially drainage grates and catch basins.

Consider the location of subsurface transit infrastructure, utilities, and sleeves to avoid conflicts with stormwater projects.

Take stock of adjacent buildings with basements or subsurface structures. Stormwater facilities may require liners or deeper walls to prevent lateral water seepage into basements.

Bioretention Planter

A bioretention cell with vertical walled sides that detains and infiltrates stormwater into the soil below.

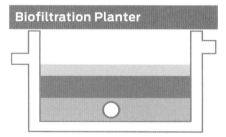

Biofiltration Planter

A biofiltration cell has vertical sides, a closed bottom, and an underdrain. This cell treats runoff for quality before draining into a perforated pipe that directs flow back into the sewer. The soil and/or an orifice control device attenuate flow to reduce peak flows into the sewer.

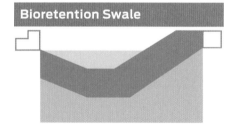

Bioretention Swale

A bioretention swale has graded side slopes, allowing greater variety of plantings and infiltrating into the soil below. It can also be designed with an underdrain pipe and lined system if infiltration is not desirable.

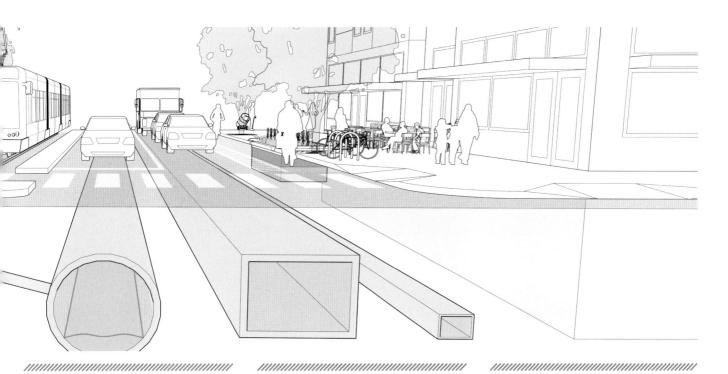

LONGITUDINAL SLOPE

PROJECT SCALE & GOALS

Consider other structures in the right-of-way such as transit stops, utility poles, mature trees, benches, wayfinding signs, or other street furniture. Maintain ADA access around doors and access routes.

Identify opportunities for synergy, especially with adjacent redevelopment or property owners interested in beautification or flood mitigation.

Streets with less than 5% slope are generally best for GSI projects, though innovative design strategies can make projects on steeper streets possible.

On steeper streets, shorter cells with frequent berms can help to slow and temporarily store runoff before infiltration.

Consider the level of stormwater management that needs to be accomplished, as set by city policy, permitting requirements, or consent order. Some projects may only need a single bioretention cell, while others may need multiple cells on multiple street blocks to manage the desired runoff volume.

If tree canopy and shade is a project goal, identify opportunities to include trees.

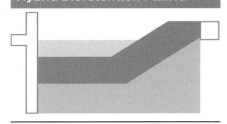

Hybrid Bioretention Planter

Stormwater Tree

Pervious Lane

A hybrid cell has one side with a vertical wall, and one graded side slope, either draining into a perforated pipe or infiltrating into native soil.

Trees for stormwater management can be planted in wells (walled cells housing a single street tree), tree trenches (linear cells with multiple spaced trees), or connected tree pits.

Permeable pavement in the roadway infiltrates runoff directly underneath, and is comprised of permeable, interlocking concrete pavers or porous asphalt or concrete. Permeable paving is suitable in contexts with lighter use.

Retrofitting Streets for Stormwater

In most cities, retrofit opportunities outnumber street reconstruction opportunities. When retrofitting a street to incorporate green stormwater infrastructure, evaluate the street with a critical eye, and look for opportunities to make geometric and operational changes.

Dean Street, **NEW YORK, NY**

REALLOCATE AVAILABLE SPACE

Assess existing features, structures, and infrastructure within the right-of-way. Review setbacks and clearances from driveways, loading zones, bus stops, utilities, buildings, mature trees, and other features.

Assess traffic volumes and parking utilization to identify streets that have excess capacity for automobiles or whose motor vehicle traffic can be served by shared surfaces. Reallocate space from overly wide streets and lanes or underutilized on-street parking to active transportation and green infrastructure.

In areas with high pedestrian activity, sidewalks typically include width for street furniture between the pedestrian zone and curb line. In residential neighborhoods with less pedestrian activity, the furnishing zone may be landscaped between the sidewalk and curb edge. Locating bioretention facilities and pervious surface where width allows can accomplish infiltration while providing opportunities for tree canopy and further buffering pedestrians from motor vehicle traffic.

PROTECT EXISTING INFRASTRUCTURE & MATURE TREES

The condition of existing infrastructure such as pavement, above and underground utilities, transit facilities, and trees may dictate the scope of a retrofit opportunity and inform construction cost estimates. For example, to avoid replacing existing utilities, further setback to protect the utilities may be required.

Street trees contribute significantly to a city's greening and stormwater management, but may pose a constraint in retrofit or reconstruction projects. Mature trees are often protected as part of a city's efforts to improve urban canopy coverage, air quality, and public health while reducing the urban heat island effect.

The recommended proximity of bioretention facilities to existing mature trees depends on the species and root structure. With planning and consideration, both systems can be accommodated and can jointly contribute to a successful green street.

SEEK COMPLEMENTARY OPPORTUNITIES

Assess the existing street design for opportunities to improve safety, pedestrian access, or transit operations. There may be bioretention strategies that can address other concerns as part of a single retrofit project. Aligning a stormwater retrofit with a mobility project is not only spatially efficient; coordination can also unlock additional funding and resources.

Test new street configurations with interim materials to evaluate the mobility impacts and human behavioral changes. Use pilot projects to experiment with retrofit designs before investing in a full street reconstruction.

Green infrastructure can contribute to placemaking initiatives, enhancing public space with greenery and other amenities. Align GSI projects and funding with placemaking initiatives to incorporate bike parking, bike share stations, street cafes, festival streets, and spaces for community gathering.

Reconstructing Streets for Stormwater

Once built, city streets are generally not reconstructed for many decades. When a reconstruction opportunity arises, it is critical to carefully consider future needs and fully capitalize on the opportunity. To extend the street's useful life, account for the expected future changes in mobility patterns and travel behavior, local climate and precipitation, and land use or development.

Full reconstruction of an existing street typically offers more flexibility for identifying opportunities for bioretention and reallocating underutilized space to bioretention areas.

University Avenue West, **ST. PAUL, MN**

COORDINATE THROUGHOUT DESIGN

Full reconstruction provides the opportunity to intentionally coordinate bioretention facilities with other infrastructure in the right-of-way, including utilities, sidewalks and curb extensions, transit stops, bikeways, bike share stations, curb cuts, and crosswalks.

Optimize the benefits of green infrastructure. Look for opportunities to co-locate bioretention facilities with other livable street design strategies, such as curb extensions or bike-lane buffers.

Consider opportunities to fully regrade the roadbed to more efficiently collect and direct runoff to bioretention facilities. Reversing the street crown to flow toward median bioretention, or grading the roadbed to slope to only one side of the street (a "thrown" street) can maximize the amount of right-of-way available to bioretention and/or provide more area to treat off-site water from upstream blocks. Regrading depends upon existing development and whether there is flexibility in adjusting grades at the property line.

ORGANIZE THE SUBSURFACE

Overhead or underground utilities, including power and communication lines, can be sited separately from the bioretention facility to maintain easy access to both infrastructure systems and prevent conflicts.

The location of service and franchise utilities is important to confirm that future connections and maintenance will not be limited by the location of the bioretention facilities. The utilities have setback and clearance requirements that will need to be accommodated during installation of bioretention. Depending upon the utility purveyor, service utilities may be able to sleeve under a bioretention facility or be required to be relocated around the facility. It is important to discuss requirements with utility purveyors during the initial project planning phase.

PLAN FOR THE FUTURE

Plan for the mobility future; design walking, bicycling, and transit facilities to encourage and accommodate growth in these modes and activity in public space, and seize opportunities to integrate multi-modal street design and sustainable stormwater infrastructure into cohesive urban design.

Install empty utility sleeves and electrical conduits under the surface, and clear of bioretention facilities, to allow for future utility connections and transportation facilities requiring power (such as transit shelters and curbside kiosks). Proactive accommodation of subsurface space can save reconstruction or retrofit costs as streets evolve over time.

University Avenue, **MINNEAPOLIS, MN**

3 Stormwater Streets

SW Lincoln Avenue, PORTLAND, OR

Stormwater Street Types

Green stormwater infrastructure (GSI) is not only a key component of a city's hardworking built infrastructure—it also acts to support social ties within a neighborhood through beautification and place-making. It can improve mobility and safety for street users, as well as visibly reassert the important hydrological and environmental functions of streets.

While both streets and cities come in many forms, GSI can be integrated into many types of stormwater streets in any urban environment. The following street types present recommendations for selecting appropriate GSI elements in selected contexts, and provide guidance for the design and placement of elements to function effectively as both hydrological and social infrastructure.

Elements are categorized based on their location in the cross-section, as well as how long they have been in common use:

- GSI elements that are well suited for and have been successfully implemented in a number of North American cities.
- GSI elements that may be implemented, but may require careful consideration given the context. Cities may still be testing these configurations.

Specific GSI element design is further detailed in Chapter 4 Stormwater Elements (page 75).

Ultra Urban Green Street

Busy urban streets play a central role in cities, and their comfort for people is a critical factor in cities' success. Often found downtown or serving as major multi-modal corridors, ultra urban streets have dense activity and strong demand for street and curb space throughout the day. While ultra urban streets are challenging to retrofit for stormwater management, the benefits of green infrastructure are greatest here: shade for public space and sidewalks, good drainage for bikeways and crosswalks, beautification at transit stops, and the capture or filtration of otherwise polluted runoff.

Integration of green stormwater infrastructure requires a high level of coordination and flexibility. Stormwater infrastructure should be designed to balance circulation, access, and mobility for diverse users and uses.

Existing Condition

EXISTING CONDITIONS

Heavily used urban streets with vast amounts of impermeable surface generate heavy stormwater runoff. Wide roadbeds and large curb radii, which encourage high vehicle speeds, result in long pedestrian crossings and unsafe walking and bicycling conditions. Street space is in high demand, both for movement and for access to each block.

Shade and wind shelter are often scarce on ultra urban streets; street trees are undersupplied and wind tunnels are common.

Curbside access is a high priority for streets with intensive commercial activity involving for-hire vehicle pick-up, freight loading and deliveries, transit and bicycle access to the curb, bike share station access, on-street parking, and accessible parking and paratransit access.

Dense utility infrastructure, duct banks, subterranean basements, and underground transit infrastructure raise the level of coordination required for GSI implementation.

RECOMMENDATIONS

Aligning GSI integration with concurrent redesign projects can leverage funding from more sources, and allows for greater coordination of planning efforts. Ahead of reconstruction, create space for GSI by testing new street configurations using interim treatments, including narrower lane widths, widened sidewalks, extended curbs at intersections, and protected bikeways. Incorporate green infrastructure during reconstruction.

1 GSI is often most effectively integrated into the planting/ furnishing zone between the curb and pedestrian through zone, where sidewalk width allows. In the illustrated example, a series of "on-line" bioretention planters infiltrate runoff, each conveying overflow to the next planter.

Busy urban streets with high vehicular and truck traffic may have greater-than-typical sediment and debris loads washed into bioretention facilities, requiring larger presettling zones at inflow points (see page 108).

Use curb extensions at intersections and midblock locations to increase visibility of and reduce risks to people crossing the street; integrate GSI into curb extensions to achieve concurrent stormwater benefits. Reducing curb radii decreases crossing distance and motor vehicle speeds, improving multi-modal safety and comfort.[1]

2 Siting GSI on lower-volume side streets may ease maintenance demands, particularly if there are concerns about people driving motor vehicles into curbside facilities.

Where GSI competes with pedestrian activity for sidewalk space, deeper bioretention planters with vertical sides maximize retention volume while minimizing the amount of sidewalk space required. Maintain a comfortable pedestrian width (typically 8–12 feet in dense contexts) when siting GSI on the sidewalk to prevent people from stepping into facilities.

Biofiltration planters that provide water quality treatment and reduce runoff volumes (page 80) are effective where water cannot be infiltrated into the sub-base, such as locations with seepage into basements, transit tunnels, or underground utility corridors.

Reconstruction

Where bioretention facilities compete for space with subsurface utilities, consider using small but frequent bioretention cells and reroute utilities for short distances. Permeable pavements may be implemented as well to reduce point-load of stormwater onto subsurface infrastructure.

As pedestrian volume increases, bioretention facilities should either be shallower or have visibly differentiated edge treatments (potentially including short curbs or low fencing) to reduce the risk of stepping in the planter. Integrate seating and placemaking elements into the bioretention facility to promote green streets as places for people.

Ultra urban streets have frequent and intensive demands on curb space. Accommodate curbside access where necessary. Maintain clear paths and avoid siting vertical elements at accessible parking spaces and designated loading zones for freight or passenger pick-up.

③ Transit bulbs or boarding islands that enable in-lane bus and rail stops and more efficient boarding may also create spaces for GSI. The illustrated "off-line" stormwater facility collects runoff from a designated street area, and conveys overflow to existing gray infrastructure. Greening at the transit stop also improves comfort for people waiting for transit.[2]

Avoid blocking sightlines between transit operators and waiting passengers; site larger trees along the back of the transit shelter or sidewalk, and use low vegetation for green facilities at the near edge of the transit stop. (See page 96 for further guidance.)

In many cities, downtown business improvement districts or similar organizations partner with cities to provide vital maintenance functions, and can be essential partners in supplementing maintenance efforts on planted bioretention facilities.

Potential GSI Features

Floating / Offset from Curb
» Bioretention Planter ○

Bikeway or Parking Lane
» Permeable Pavement ●

Planting Zone
» Tree Well or Trench ●
» Bioretention Planter ●
» Bioretention Swale ◔

Sidewalk
» Permeable Pavement ◔

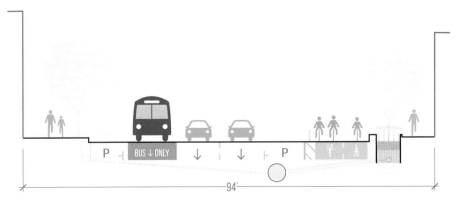

Case Study: Fell & Oak Streets

Location: San Francisco, CA

Street Context: Urban Corridor

Project Length: 0.5 miles

Right-of-Way Width: 65 feet

Participating Agencies: San Francisco Municipal Transportation Agency (SFMTA), San Francisco Public Utilities Commission (SFPUC), San Francisco Department of Public Works (SFDPW)

Timeline: *Interim design,* October 2012–May 2013 *Permanent design completed* May 2015

GOALS

Stormwater management: Manage a 5-year/3-hour storm event, or 1.25 inches of rain.

Mobility: Improve safety and comfort for people walking and biking along Oak and Fell Streets.

Placemaking: Improve aesthetics of the streets in the project area.

Oak Street bikeway, **SAN FRANCISCO, CA**

OVERVIEW

Fell and Oak Streets are vital links in San Francisco's bicycle network, but featured conventional, unbuffered bike lanes next to relatively fast-moving traffic, resulting in an uncomfortable biking experience. The streets were redesigned with protected bike lanes separated from motor vehicle traffic by flex posts, bulbouts, daylighted intersections, and enhanced crosswalks.

The Fell and Oak Streets area was later upgraded to integrate stormwater management features, including bioretention planters in expanded and upgraded corner bulbouts, permeable pavers to channel rainwater along the gutter, and planters in the bikeway buffer.

DESIGN DETAILS

The project integrated rain gardens into corner bulbouts, reclaiming space for stormwater management while simultaneously shortening crossing distances, improving pedestrian safety, and enhancing the street's aesthetic value. The bulbouts capture runoff from the sidewalks.

A 5.5-foot-wide strip of permeable pavement in the parking lane adjacent to the curb absorbs storm flows from the street.

The bikeway is separated from motor vehicle traffic by a 4-foot-wide concrete median that holds planter boxes to catch rainwater and beautify the corridor.

LESSONS LEARNED

Coordinate early and often. The original project team, SFMTA and SFDPW, received community requests for green infrastructure during project's outreach phase. The stormwater design team at SFPUC was brought into a project already in process, requiring all stakeholders to play catch-up and learn to coordinate while already in process.

Take advantage of existing conditions. A primary lesson that SFPUC learned during this project, a test case for future projects, is to collect quality survey and existing conditions data, especially of local infiltration capacity. By capturing the infiltration of native soils, designers are able to correctly size and model facilities, saving on design and construction costs while meeting stormwater goals.

Don't sacrifice "green for green." Many locations in San Francisco have to balance between a need for improved tree canopy as well as bioretention.

Demonstrate the value of investments. Projects with unexpected cost increases could make it more difficult to justify spending on improvements later on. Be sure to accurately measure costs and projected outcomes to pick projects and designs that offer the most value.

OUTCOMES

Preliminary measurements indicate that the project is managing 90% of the stormwater runoff from the street.

98% of riders surveyed felt that the safety of bicycling in the project area had increased.

Fell Street, **SAN FRANCISCO, CA**

Oak Street, **SAN FRANCISCO, CA**

Fell Street, **SAN FRANCISCO, CA**

Fell Street, **SAN FRANCISCO, CA**

Green Transitway

Upgrading to a high-capacity rail or bus transit line within the street often involves significant capital investment or full reconstruction, creating an opportunity to reframe the entire form and use of the street. Green stormwater infrastructure is integrated comprehensively in this street, recasting a large impervious expanse of pavement into a high-performing stormwater street that provides an attractive and safer street for people, both on transit and off.

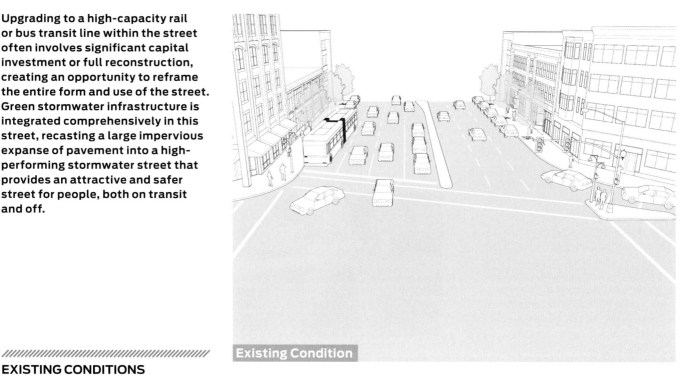

Existing Condition

EXISTING CONDITIONS

Major thoroughfares that serve transit connecting the city's downtown and neighborhoods may have significant imbalance toward motor vehicle capacity that is poorly used at most times of day, creating daunting or intolerable walking and bicycling conditions, and broad swaths of impermeable surface that exacerbate stormwater runoff and outfall events. Auto-oriented land uses along major streets create polluted runoff, making water quality treatment a priority.

These wide streets are often designed for peak-hour automobile volume, are underutilized throughout the rest of the day, and have not been optimized for transit. Overprovision of motor vehicle space results in frequent speeding and poor walking and biking conditions, degrading the public realm and depressing adjacent property values and land uses that would generate activity.

Moderate-to-heavy turn volumes and an overly high design speed result in long turn bays and wide curb radii, covering large areas of the right-of-way with impermeable surfaces. Left turns across multiple lanes introduce high-risk conflicts between motor vehicles and other street users.

RECOMMENDATIONS

1 High-capacity transit can move more people in less space, unlocking street space for GSI and improving environmental performance. Shifting trips from cars to transit and lowering motor vehicle speeds has the effect of decreasing emissions, surface pollutants, and noise impacts while increasing safety. Use newly identified space, including medians, curb extensions, and bikeway buffers, to provide high-capacity stormwater facilities.

Dedicated right-of-way for high-capacity travel modes may itself be integrated with stormwater facilities; rail guideways or concrete busways involve large continuous linear spaces, which can be outfitted to infiltrate stormwater without interrupting accessible paths. As illustrated, linear bioretention planters capture stormwater as sheet flow (unchannelized runoff) from the transitway. Curbs are not necessary alongside fixed guideways, but are recommended next to motor vehicle or bicycle travel lanes to reduce incursion.

Wide rights-of-way with less dense underground utilities sometimes present greater opportunities for stormwater infiltration than smaller streets. Long blocks with few curbside access needs and wide planting strips may allow for implementation of large swales with significant infiltration capacity. Plantings and trees not only improve stormwater performance, but also help to transform auto-oriented streets into places for people, absorbing noise and air pollution and introducing valuable habitat into a corridor.

2 The right turn lane has been replaced with a green refuge island, and turn speeds have been reduced with a tighter curb raidus. The left turn has been restricted or given a signal phase to improve safety and transit reliability. The left turn lane has been closed and repurposed as part of the green median, with a pedestrian refuge island.

3 The slip lane, which had enabled unsafe high-speed turns, is closed and its footprint reallocated to a large bioretention planter. A pedestrian walkway is provided along the diagonal desire line, creating an opportunity for informational signage and placemaking.

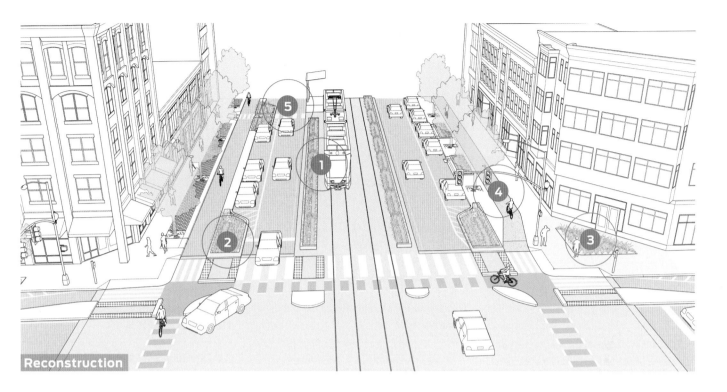

Reconstruction

As speeds decrease, the risk of vehicles entering stormwater facilities also decreases. Reduce target speed to safe and appropriate urban speeds—typically 20–25 mph, and rarely more than 30 mph—through the reassignment and narrowing of motor vehicle traffic lanes; a transit-supportive signal progression speed, usually below 20 mph; and traffic calming elements. High-visibility or retro-reflective vertical elements may be placed at the leading edge of GSI facilities to reduce the risk of motor vehicle incursion, especially where car speeds remain high.

4 Protected bike lanes and raised cycle tracks provide safer and more-comfortable bicycling conditions for people of all ages and abilities, distinct and separate from both the sidewalk and motor vehicles. Permeable pavement may be used on the bikeway: use pervious concrete or porous asphalt to ensure the surface is compatible and comfortable for bicyclists. If at street grade, runoff from the street can be directed into the permeable bikeway; if raised, the permeable bikeway can absorb sidewalk sheet flow. Plan regular maintenance work to ensure the permeable pavement does not get clogged with sediment.

5 Midblock signalized crossings are opportunity sites for bioretention facilities, such as midblock curb extensions. Plantings must be low to maintain sight lines, especially just upstream of the crossing. Provide accessible paths and ramps to the crossing with vertical or other detectable elements separating paths from the stormwater facility.

Potential GSI Features

Median Transitway

» Bioretention Planter ●

» Bioretention Swale ○

Planting Zone

» Tree Well or Trench ●

» Bioretention Planter ●

» Bioretention Swale ○

Floating / Offset from Curb

» Bioretention Planter ○

Bikeway or Parking Lane

» Permeable Pavement ●

Sidewalk

» Permeable Pavement ○

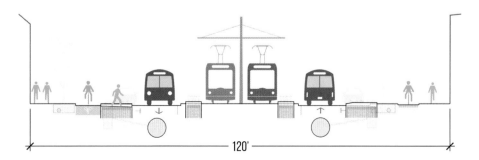

120'

Case Study: METRO Green Line

Location: Minneapolis & St. Paul, MN

Street Context: Green Transitway

Project Length: 11 miles

Right-of-Way Width: 100 feet

Participating Agencies: Metro Transit, Metropolitan Council, Ramsey County, City of Minneapolis & City of Saint Paul

Timeline: *Planning started* 2001
Construction, 2011–2012
Transit service started 2014

Cost: $957 million, including $5 million for green investments

GOALS

Mobility: Provide high-quality, regular, and reliable transit service connecting Minneapolis and St. Paul.

Stormwater management: Reduce stormwater runoff and improve water quality.

Economic development: Improve connectivity between Minneapolis and St. Paul, enhancing access to a thriving and vibrant corridor.

University Avenue West, **ST. PAUL, MN**

OVERVIEW

The METRO Green Line is an eleven-mile light rail route connecting downtown Minneapolis and downtown St. Paul, two adjoining cities that together make the core of a metropolitan region. The route cuts through highly developed, large tracts of industrial, commercial, and residential areas and links five major centers of activity, including the University of Minnesota.

University Avenue, a significant regional transportation corridor, was fully reconstructed to accommodate the light rail line, providing the opportunity to rethink the corridor's ability to capture and manager stormwater runoff. Significant investment in green stormwater infrastructure leverages the local hydrology to treat stormwater through infiltration, reducing stormwater pollution reaching the Mississippi River.

DESIGN DETAILS

Rain gardens, bioretention planters, infiltration trenches, permeable pavers and tree trenches work along the corridor and on side streets to absorb and filter stormwater. Public art is integrated into the design of infiltration trenches and transit stations along the corridor, creating a sense of place.

Tree trenches, with over 1,000 planted trees, line five miles of University Avenue. Stormwater runoff enters the trenches through permeable pavers and surface catch basins, and is then stored in rock-filled cells. Planted trees along the trenches absorb and filter stormwater, and stormwater is slowly released into the underlying soil. In the case of extreme rain events, excess stormwater is carried into the storm sewer.

Infiltration trenches run underneath two side streets in St. Paul. Stormwater flows through storm drains into large pipes under the street. These pipes, punched with thousands of holes, are buried in long trenches filled with golf-ball-sized rocks. The water fills the pipes and a controlled release through the holes allows the water to infiltrate the rocks and soil. Excess stormwater from extreme rain events is carried into the storm sewer.

University Avenue West, **ST. PAUL, MN**

Bioretention planters were implemented on five side streets. Curb openings channel stormwater from the roadway into the planter, where hardy native plants, modified soil, and sand filter stormwater while recharging the ground reservoir.

Rain gardens were built on four adjacent streets along the rail corridor, improving water quality in the project area.

Permeable pavers were applied as the surface treatment above the tree trenches, providing for infiltration of rain water from the sidewalk as well as improved air exchange between the surface and the soil.

KEYS TO SUCCESS

Unique partnership. Between 2007 and 2016, the Central Corridor Funders Collaborative was formed to align interests and identify potential housing, economic development, and greening opportunities along the Green Line. The Funders Collaborative worked to complement the work of the municipal agencies. The members have made more than 160 grants, totaling nearly $12 million dollars of investment.

Leverage multiple funding sources. The cost of the green infrastructure portion of the project leveraged numerous funding sources at multiple levels of government, including a state Clean Water Fund grant, regional funding from the Metropolitan Council, local funding from the City of St. Paul, and in-kind contributions from the Capital Region Watershed District.

Monitor progress. Monitoring equipment was installed to assess the bioretention facilities' performance, and results are expected to inform future green infrastructure investments.

LESSONS LEARNED

Align utility infrastructure installation. Redundant efforts and extended timelines can be reduced by closely coordinating activities.

Meet early and often with stakeholders. Stakeholders can positively influence the outcome of a project, and potential conflicts can be addressed early. Many design elements were influenced through stakeholder outreach, including new curbs and gutters to improve the stormwater elements of the project.

Invite participation. Work with unique partnerships, art organizations, neighborhood groups, and local businesses to enrich and fine-tune project goals to meet the needs of the community.

OUTCOMES

Approximately 50% of stormwater runoff has been mitigated by the green infrastructure.

80 pounds of phosphorous and 40,000 pounds of sediment are treated and removed by the green infrastructure annually.

The project utilized materials with a high recycled content, including over 30% recycled steel for the 12,000 tons of track rail installed.

Over 1 million transit riders a month enjoy service between the two cities. Passengers can access the Green Line at 18 stations along the route.

The neighborhoods along the Green Line have experienced more than $3 billion in commercial and residential development.

13,700 housing units have been added or planned within one-half mile of the line.

SE Washington Avenue, **MINNEAPOLIS, MN**

Boulevard

Large boulevards with multiple wide roadbeds are often overbuilt for vehicle traffic. Many cities are restoring these boulevards into vibrant, iconic urban streets, with increasing opportunities to increase safety and comfort for people using the street, capture stormwater runoff, improve shade coverage, and reduce reflective surfaces. These ecosystem services are key to making broad streets into good human habitats.

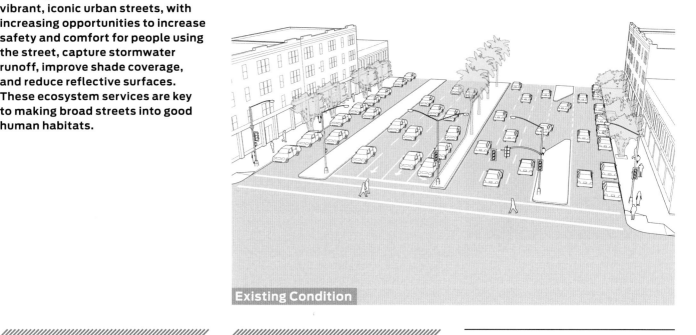

Existing Condition

EXISTING CONDITIONS

Urban boulevards have historically been designed to maximize automobile throughput, often resulting in poor conditions for people walking, biking, and riding transit. Like other wide arterials, they are prone to frequent speeding and unsafe vehicle operations, but are vital routes for all modes.

Land use and destinations are often mixed and dynamic, with many different types of users and visitors. Both through movement and destination access are in high demand.

Boulevards include broad swaths of impermeable surface that exacerbate stormwater runoff and outfall events. A wide right-of-way with multiple roadbeds creates a complex context for managing a significant volume of stormwater runoff. During heavy rainfall, the service roads—where most walking and bicycling activity takes place—are at particular risk for fast-moving runoff streams and pools.

RECOMMENDATIONS

Boulevards provide some of the greatest opportunities for green stormwater infrastructure in cities due to their large impermeable footprints and extraneous motor vehicle areas. Service road medians and curbsides provide significant space for water retention, storage, and treatment.

1 Mature trees—which perform significant stormwater management—and historic or unique urban design can be found on many boulevards, and should be used to anchor public space and placemaking opportunities. Integrate water management into the aesthetic of the street to create high-quality, comfortable places for people while designing to protect and promote the root structures of mature trees.

Choose a low target speed and design the entire street around that speed in order to reduce space consumption by motor vehicles, especially at lateral shifts. Compact slip lanes that enforce very low-speed access to the service roads unlock additional pedestrian and green infrastructure space.

2 This left turn lane has been shortened and signalized, and the median has been extended as a refuge. Boulevards require careful design at intersections with cross traffic. Enable pedestrians to make safe, easy crossings at desire lines at each intersection, and at least every 300–400 feet on long blocks. Manage or restrict turns from the center roadbed to eliminate unsafe turning conflicts.

3 Transit stops and stations, especially on the side median, are ideal siting opportunities for bioretention facilities, and may be continued as linear pedestrian spaces to improve pedestrian access and comfort.

Though center median facilities provide large linear footprints, they may be more expensive places to implement bioretention facilities, as they are often difficult to access for maintenance and may require reversal of the roadway cross slope to redirect stormwater runoff to the median. Assess costs of siting bioretention facilities in the median in context with potential benefits, and consider whether street trees are better suited to that space, which still improve shade canopy and perform some infiltration.

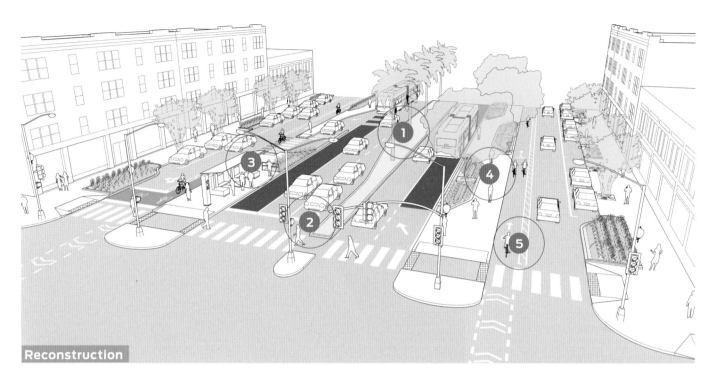

Reconstruction

4 Where space is available, bioretention swales with graded side slopes (see page 82) may allow gentler transitions from the pedestrian path to the bioretention facility and allow wider diversity in plant selection. The bioretention facilities can also provide a buffer between the pedestrian path and motor vehicle traffic, especially along the side median.

Local deliveries access abutting businesses on the service roads. Freight access management is an important tool to mitigate safety conflicts on multi-way boulevards. The center road beds of boulevards are often key truck routes, and are sometimes designated as such; 11-foot right lanes on the center roadway are often appropriate , though other lanes should be 10'. The service roads should be distinct in design to discourage trucks from using them as through routes.

5 Bike facilities for all ages and abilities should be prioritized along boulevards and other continuous major streets due to the available width and demand for local and regional bicycle traffic. Permeable pavement may be used for the bike route as long as run-on from adjacent impervious areas is minimized. Utilize pervious concrete or porous asphalt to ensure the surface is compatible and comfortable for bicyclists.

Boulevards with high vehicular use and truck traffic will generate large sediment and debris loads depositing into bioretention facilities, requiring larger-than-typical presettling zones at inflow points.

Potential GSI Features

Median/Pedestrian Boulevard
» Bioretention Swale

Bikeway & Parking Lane
» Permeable Pavement ●

Sidewalk Planting Zone
» Bioretention Swale
» Bioretention Planter ●
» Tree Well or Trench ●

Curb Extensions (corner or midblock)
» Bioretention Planter ●
» Bioretention Swale

Sidewalk
» Permeable Pavement

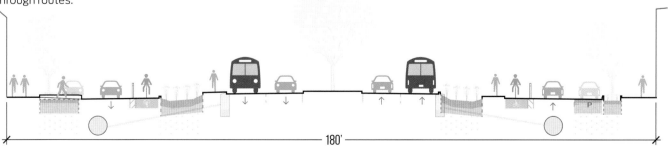

180'

Case Study: 28th/31st Avenue Connector

Location: Nashville, TN

Street Context: Multi-way Boulevard

Project Length: 0.3 miles

Right-of-Way Width: 88 feet

Participating Agencies: Metro Nashville Public Works, Metro Water Services, Metro Arts Commission

Timeline: 2011–2012

Cost: $6.3 million

GOALS

Connectivity: Reconnect two city neighborhoods: North Nashville and West End.

Mobility: Improve traffic flow and transit service between six area universities and medical centers.

Demonstrate possibilities: Serve as a high-profile example of complete streets design in Nashville.

OVERVIEW

The 28th/31st Avenue Connector reconnected neighborhoods that had been separated for decades by railroad tracks and a major interstate. Given the high visibility of the project, called for by community members and city leaders for years, Nashville decided to design the project as one of the city's first complete streets projects, along with substantial green stormwater infrastructure.

28th/31st Avenue Connector, **NASHVILLE, TN**

The 28th/31st Avenue Connector provides a bridge crossing over an existing railroad to connect West and North Nashville. The reconstructed street is one of the city's early high-profile complete street projects, providing safe and comfortable accommodations for all users. Additionally, the connection was an early demonstration of "green street" design strategies, integrating bioretention facilities and improved human environment.

Designed from the ground up, the project incorporates physically separated bike lanes and sidewalks, attractive landscape design with native plant species and shade trees, along with embedded LED delineator lights between bike and pedestrian walkways to illuminate and visually separate the pathways while minimizing environmental impact.

DESIGN DETAILS

The reconstructed green street is meant to reduce stormwater runoff volume and to provide water quality treatment. Median and curbside bioretention cells are designed to capture the "first flush," treating the heaviest pollutant loading that washes off the road during storm events. The street is constructed to slope in one direction rather than with a crown, directing all runoff toward green infrastructure from the top of the bridge down in either direction.

The project includes narrowed travel lanes to reduce impervious asphalt surfaces. Curb cuts direct stormwater to bioretention planters and swales, filtering runoff through planting and engineered soil layers. Every 30 feet, concrete dams slow stormwater flow, decreasing runoff and increasing the biorentention capacity of the project.

Median bioretention swales are constructed up to 10 feet wide, and curbside swales are 4 feet wide behind the curb. The curbside bioretention cells are designed with a maximum 4-inch depth below the sidewalk and bikeway level, which provides adequate retention depth to manage up to a 100-year storm. Low plantings discourage people bicycling and walking from accidentally entering the bioretention swales, while the relatively shallow ponding depth provides a much gentler transition in the case of accidental entry.

The bicycle facilities are paved with dyed concrete to differentiate them from pedestrian paths on either side of the street. Embedded LED lights provide additional definition between modal spaces and improve comfort for people bicycling and walking.

Local artist-designed transit shelter,
NASHVILLE, TN

KEYS TO SUCCESS

Do high-profile projects right. The 28th/31st Avenue Connector is one of Nashville's first complete streets, and is credited with putting complete streets at the center of city infrastructure discussions.

Closely involve all stakeholders. Monthly stakeholder meetings ensured a high-quality outcome, and minimized disruptions for adjacent businesses, including a large regional hospital.

Design for future needs. The connector was designed to accommodate current and future development demand, as economic and social activity increases between the two reconnected neighborhoods.

Create a sense of place. Community suggestions were transformed into large-scale public art, with quilt panels covering a safety fence, along with six artist-designed bus shelters, giving the project area a clear sense of identity and place.

OUTCOMES

Reconnected two neighborhoods long bisected by a railroad and a highway.

Enabled a new transit route and service that was not logistically feasible until this project.

Provided a high-profile complete street for Nashville, prominently showcasing high-quality bike and pedestrian infrastructure.

The 28/31st Avenue Connector project influenced the city's approach to future projects. A Complete Streets executive order requires all new projects to include bioretention elements such as walled planters and graded swales.

28th-31st Avenue Connector, **NASHVILLE, TN**

Neighborhood Main Street

Neighborhood main streets are at the center of community life. They are conduits for social and economic activity, providing vital spaces to travel to and through.

Green expressions make streets more inviting by providing a tree canopy, absorbing heat, and improving the image of the street. Effective stormwater management is critical to mitigate flooding risks for adjacent properties on neighborhood main streets.

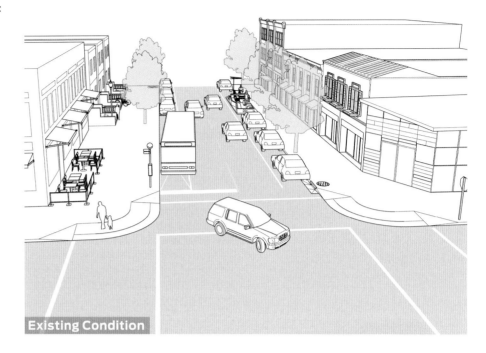

Existing Condition

EXISTING CONDITIONS

Less dense than downtowns, neighborhood main streets serve local business activity and civic life, and are characterized by high demand for a quality walking and bicycling environment, frequent parking turnover and freight access, and service by key transit routes.

Adjacent property owners rely on effective stormwater management in the right-of-way to prevent damage caused by basement or building flooding.

RECOMMENDATIONS

Green expressions—including bioretention facilities, street trees, and landscaping—enhance neighborhood main streets, creating more aesthetically pleasing public spaces even where the street is relatively narrow.

Road diets and the reassignment of road space to bicycling and walking can improve mobility and safety by moving more people in less space and reducing left turn and rear-end conflicts among private motor vehicles and between motor vehicles, bicycles, and transit vehicles, while simultaneously unlocking street space for stormwater treatment facilities.

1 Curb extensions with bioretention facilities should be integrated at intersections and midblock locations to improve pedestrian mobility and safety, shorten crossing distances, and calm vehicle traffic by narrowing the road. Bioretention cells can be sited throughout the block, though are most effective when sited near intersection corners to collect as much runoff from the tributary area as possible.

2 Transit boarding bulbs are an important opportunity to integrate green infrastructure since sidewalk space is often not available and curbsides are at a premium.

3 Smaller green infrastructure treatments, such as bioretention planters, stormwater tree wells, or tree trenches, are more commonly used on neighborhood main streets due to space constraints and high foot traffic along the sidewalk and between the curb and storefronts.

The bioretention facility wall can incorporate seating and placemaking elements in the planting or furnishing zone, especially on main streets with significant foot traffic and active storefronts.

On busy pedestrian corridors, consider edge treatments around bioretention facilities, such as short fencing or dense plantings, to reduce the risk of people stepping in the planter.

4 Successful and well-supported interim treatments such as parklets may be incorporated into capital reconstruction, and can create sites for bioretention and infiltration.

Reconstruction

Where local business activity generates heavy freight and delivery activity, GSI elements should be located and designed for compatibility with frequent curbside access. Site bioretention planters to allow for access to the sidewalk, including regular spacing and accessible paths through the planting strip. Consider reducing parking duration, shifting deliveries to off-peak hours, designating freight loading zones during specified hours, or siting designated freight loading zones around the corner on side streets to reduce conflicts and facilitate local business activity.

At food, entertainment, and retail destinations, collaboration with local business is often key to a successful GSI project. On these highly active sidewalks, more frequent cleaning of litter and debris from bioretention facilities should be expected. The aesthetics of stormwater facilities are often more important than functionality to abutting businesses and street users. Consider enacting maintenance agreements for debris removal and light weeding with specific businesses, business improvement districts, or merchant organizations to reduce costs and improve performance.

Bioretention facilities may need to be lined to prevent groundwater migration into adjacent structures, such as basements or utility duct banks.

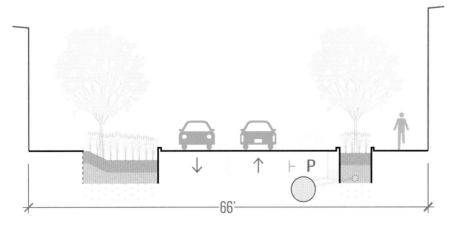

Potential GSI Features

Bikeway / Parking Lane

» Permeable Pavement ●

Curb Extensions (corner or midblock)

» Bioretention Planter ●

» Bioretention Swale ○

Sidewalk Planting Zone

» Bioretention Planter ●

» Bioretention Swale ○

» Tree Well or Trench ●

Case Study: Southeast Division Street

Location: Portland, OR

Street Context: Neighborhood Main

Project Length: 1.6 miles

Right-of-Way Width: 60 feet

Participating Agencies: Portland Bureau of Environmental Services, Portland Bureau of Transportation

Timeline: 2009–2015

Cost: $13 million, including $2.5 million in federal transportation funds and $3.3 million in local transportation funds

GOALS

Stormwater management: Eliminate sewer backflows into basements and streets in the project area.

Water quality: Reduce the severity and frequency of combined sewer overflow events, meeting EPA guidelines for discharges to the Willamette River.

Mobility: Improve safety and comfort for people walking and biking along the corridor. Increase reliability and running speed of the heavily used bus lines along the street.

Access: Increase access to local businesses by providing adequate on-street parking for cars and bikes.

Placemaking: Improve aesthetics of the corridor with landscaped stormwater facilities, shade trees, and public art.

SE Division Street, **PORTLAND, OR**

OVERVIEW

SE Division Street was a deficient transportation corridor with varying land use and many competing needs. Historically, the corridor has served as a through street, serving much of the traffic traveling through surrounding neighborhoods. Over time, however, the street became a destination in itself, with more neighborhood retail shops, increasing the number of pedestrians and bicyclists along the corridor and making parking for pedestrians and cars scarce.

Safety along the corridor was lacking, with the 85th percentile for vehicle speeds ranging from 26-31 mph despite a posted 25-mph speed limit. Peak travel lanes, which replaced curbside parking during rush hours, were underutilized, with 35% of morning rush hour traffic using the peak travel lane, and 30% of morning rush hour traffic using the reverse peak travel lane.

Through an in-depth, multi-year design and outreach process, the City of Portland worked with a technical advisory group and a community advisory group on a plan for improving the corridor and accommodating growth over the next 20 years. The street was redesigned to include fewer but better-performing travel lanes, along with curb extensions, marked crossings, improved bus stops, on-street bike parking, bioretention planters to manage stormwater runoff, and large canopy trees along the corridor to provide shade and improve water quality.

Midblock crossing, **PORTLAND, OR**

DESIGN DETAILS

The project eliminated peak-hour travel lanes along most of the corridor after finding that they were underutilized. By eliminating lanes, designers reclaimed space for other uses.

A curb-to-curb reconstruction of the corridor brought sidewalks up to standard width and added or redesigned corner ramps to make the street ADA compliant.

Large canopy trees were planted on the north side of Division Street throughout the corridor, avoiding utility conflicts on the south side. The trees help shade the street, intercept stormwater, and add attractive landscaping and traffic calming elements to the corridor.

Curb extensions with integrated stormwater facilities (often also extending into furnishing zones) manage excess stormwater runoff from the street and also improve pedestrian safety by reducing crossing distance and enhancing visibility of pedestrians.

Curb extension, **PORTLAND, OR**

In some parts of the corridor, designers integrated stormwater facilities into furnishing zones to minimize impervious area while avoiding utility conflicts.

The project added on-street bike parking (bike corrals) throughout the corridor. By eliminating peak-hour travel lanes, the project also added additional on-street car parking along most of the corridor. In addition to enhancing the vitality of the retail landscape, bike and car parking add a buffer between moving traffic and pedestrians on the sidewalk.

KEYS TO SUCCESS

Involve the community from the start. Division Street's successful redesign plan drew upon over ten years of community input, consultation, and leadership, starting from a 2003-2005 vision of the corridor as a neighborhood main street, to the eventual street reconstruction plan designed from 2009 to 2012.

Adapt design to local context. While peak-hour travel lanes were removed along the majority of the corridor, additional lane capacity was warranted at three key intersections. In these targeted areas, the project included signal timing changes, bike boxes, and removal of parking to reduce conflicts with transit.

Monitor impacts. Portland is monitoring traffic flow impacts resulting from changes to the corridor, particularly to ensure that car traffic was not inadvertently diverted to a nearby street with heavy bike ridership.

Maintain business access. Stormwater facilities were carefully sited so as to not overly impact the street frontage of any one business. In many cases, facilities were 'split' between two storefronts, balancing demand for access to local retail.

LESSONS LEARNED

Avoid unclear street uses. Site stormwater infrastructure so that it, or a curb extension, extends to the corner, avoiding inadvertently creating a 'partial parking space' that drivers may attempt to park their cars in.

OUTCOMES

The Division Street project diverts 5.7 million gallons of stormwater from the combined sewer system annually.

Dense housing construction has boomed along the Division Street corridor, and an influx of retailers has made the street one of the most popular in the city of Portland for shopping and dining.

While Division Street is one of the top 30 streets for crashes in Portland, the project area has relatively few crashes compared to the rest of the corridor.

T-intersection curb extension, **PORTLAND, OR**

Transit stop, **PORTLAND, OR**

Residential Street

Local residential streets are often underutilized as public spaces, with overly wide or undifferentiated lanes that enable speeding and cut-through traffic. Stormwater projects are often the only opportunity to reconstruct or fully retrofit such streets. Integrating green stormwater infrastructure can help create calmer streets for people walking, biking, and enjoying their neighborhood.

Existing Condition

EXISTING CONDITIONS

Neighborhood streets are small, and typically have low volumes of through movement in any mode. They are sensitive to their traffic network role, and can quickly degrade if made attractive as a cut-through for motor vehicles. Many such streets are already traffic calmed, often with low-cost measures such as speed humps.

Some neighborhood streets with narrow roadbeds are configured as yield streets, permitting parking on one or both sides, with narrow travel space enforcing low speeds on motor vehicle traffic. Other neighborhood streets are marked with a lane or lanes for through traffic along with on-street parking spaces. Traffic operations are sometimes one-way. Speed is a frequent complaint.

Mature trees and frequent driveways often interrupt the sidewalk, and sometimes limit the ability to site bioretention facilities.

RECOMMENDATIONS

Design local streets for very comfortable walking and bicycling for all ages, while integrating green stormwater infrastructure. These streets serve a multitude of short trips and present a significant opportunity to integrate stormwater management into the right-of-way.

Due to low traffic volumes and less sediment and debris, neighborhood streets can be ideal right-of-way sites for bioretention facilities and/ or permeable pavements. Water from higher traffic cross streets can be conveyed to bioretention facilities on neighborhood streets, which may simplify maintenance access.

1 The planting strip creates opportunities for large infiltrating surface area; graded bioretention cells may offer a softer urban design adjacent to the sidewalk. Prioritize maintaining mature trees where possible. Consider the use of tree wells and trenches if space is constrained.

2 Designate the reconstructed street as a bike boulevard, with design strategies to manage motor vehicle speed and volume. A curb extension planter at the downstream end of the block serves as a partial closure to manage motor vehicle volume.

Curb extensions are also proven effective at increasing pedestrian visibility, shortening crossing distance, and calming motor vehicle traffic by enforcing low speed turns and through movements. Bioretention planters sited at curb extensions should be planted with low shrubs and vegetation that maximize visibility.

3 Midblock curb extensions are configured to "chicane" the street, managing yield interactions between two-way vehicle traffic and maintaining slow operations. On the narrowest streets, parking and plantings are placed on alternating sides of the street with a clear zone reserved in the center for two-way emergency vehicle access.

Reconstruction

④ Permeable pavement may be used on the full roadbed to manage runoff. In this illustration, permeable pavement under the parking lane is used to capture runoff from the street, with additional flow being directed to curbside infiltration facilities. If permeable pavement is only used in a partial zone, install a vertical liner between the two zones to protect the conventional pavement material. In cases of reconstruction, apply permeable pavement across the full roadway width to minimize maintenance needs.

Depending upon right-of-way width and available space, graded bioretention facilities may be feasible in existing planting strips and/or at curb extensions, as long as minimum bottom width can be provided for maintenance crews. In constrained spaces or in areas with high pedestrian activity, bioretention facilities with vertical walls may be more suitable.

Bioretention swale on a residential street, **SEATTLE, WA**

Potential GSI Features

Bikeway & Parking Lane

» Permeable Pavement ●

Curb Extensions (corner or midblock)

» Bioretention Planter ●

» Bioretention Swale ●

Sidewalk Planting Zone

» Bioretention Swale ●

» Bioretention Planter ●

» Tree Well or Trench ●

Sidewalk

» Permeable Pavement ●

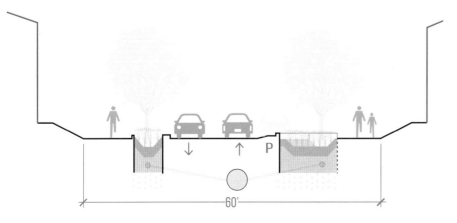

Case Study: Barton CSO Control with Roadside Rain Gardens Retrofit

Location: Seattle, WA

Street Context: Residential

Project Area: 15 city blocks

Contributing Drainage Area: 32 acres

Right-of-Way Width: 60 feet

Participating Agencies: King County Wastewater Treatment Division

Timeline: *Planning started* 2009 *Construction,* 2013–2015

Cost: $5.1 million

GOALS:

Stormwater management: Reduce combined sewer overflow events by at least 75%.

Demonstrate possibilities: Serve as a model for future green infrastructure retrofit projects for addressing combined sewer overflows.

OVERVIEW

The Barton CSO Control project is designed to address combined sewer overflows in the Barton Combined Sewer Basin from discharging into Puget Sound at an outfall near Lincoln Park in Seattle, a popular recreation area.

34th & Cloverdale, before, **SEATTLE, WA**

34th & Cloverdale, after, **SEATTLE, WA**

The Barton Street Pump Station historically had an average of four overflows per year, discharging four million gallons of polluted runoff. King County needed to reduce overflows to no more than one per year on a 20-year moving average in order to comply with the State of Washington's Department of Ecology requirements. The plan to do so includes both gray infrastructure (pump station upgrades) and green infrastructure approaches along City of Seattle streets within the combined sewer basin.

DESIGN DETAILS

King County constructed 91 bioretention cells with graded side slopes (see page 82) in the planter strip on 15 blocks in the project area. When it rains, stormwater filters through the bioretention facility soil to a slotted drain pipe, which takes the water to a deep well for slow infiltration underground. The project team determined the number of needed bioretention facilities through hydrologic/hydraulic modeling.

The design team sited bioretention facilities on residential, non-arterial streets that were relatively flat (under 5% grade), with few driveways. Sites were also selected to avoid public and private utility conflicts, have planting strips at least 9 feet wide, and have minimal impact to on-street parking and existing mature trees.

KEYS TO SUCCESS

Outline maintenance responsibilities from the onset. King County is responsible for all maintenance of the bioretention facilities, including monitoring plant health, ensuring curb inlets into the cells are clear, removing weeds and debris, and monitoring drainage. Unmaintained bioretention facilities will become ineffective over time. In addition, a level of acceptable appearance is critical to preserving the public's trust and approval of the project.

Clearly communicate community benefits. Extensive community outreach, and efforts to minimize impacts from construction, resulted in support of the project from community members. Showing community members images of what the bioretention facilities look like at all stages of construction, and at different times of year, helped manage community expectations and build enthusiasm from local stakeholders.

Allow time for the first installations. The project included initial review of the cell grading and mock-ups of key elements prior to constructing all the blocks. While this took some time, once they were approved and expectations established, the construction of the streets went smoothly and quickly.

LESSONS LEARNED

Avoid jargon. Use accessible, consistent terms to help community members understand the need for green infrastructure and its benefits. Using the term "roadside rain garden" instead of "bioretention facility" increased comprehension of and acceptance of the project.

OUTCOMES

Preliminary monitoring and observations indicate that the rain gardens with the deep infiltration wells are performing as intended, and are infiltrating all the water that is draining into the facilities.

Case Study: Pinehurst Green Grid

Pinehurst Green Grid, **SEATTLE, WA**

Biorention swale, **SEATTLE, WA**

Location: Seattle, WA

Street Context: Residential

Project Area: 49 acres / 12 city blocks

Contributing Drainage Area: 2.3 acres

Right of Way Width: 60 feet

Participating Agencies: Seattle Public Utilities

Timeline:
Preliminary Engineering, 2003–2004
Design, 2004–2005
Construction & Landscaping, 2005–2007

Cost: $4.6 million (construction cost: $2.71 million)

GOALS:

Stormwater management: Manage stormwater runoff volume to a minimum of a 6-month storm event (1.08 inches in 24 hours), up to a 2-year storm event (1.68 inches in 24 hours). Eliminate spot flooding problems and provide local drainage conveyance.

Water quality: Meet the City of Seattle and Washington Department of Ecology water quality standards for the total drainage area.

Placemaking: Improve the Pinehurst neighborhood with attractive street paving, sidewalks, and landscaping.

OVERVIEW

The Pinehurst Green Grid is a large-scale neighborhood natural drainage system treating a 49-acre drainage basin. The basin drains into Thornton Creek, a salmon-bearing urban creek. Located in a part of the city with unimproved right-of-ways, many streets had no curbs, no formal drainage infrastructure, and limited sidewalks.

The project re-aligned the road surface to one side of the right-of-way, adding a sidewalk on one side and large bioretention cells along the other side of the street. Parking was normalized to be restricted to one side of the street throughout the project area.

LESSONS LEARNED

Look beyond the project area. The large bioretention cells installed in this project can handle runoff not only from the adjacent street and houses but from an area three-to-five times that size. The additional capacity allowed for a system that would accept runoff from cross streets and other nearby streets that therefore did not need reconstruction for bioretention facilities.

Take efficiency into account. The design of the Pinehurst Green Grid was more cost-effective than a similar project, Seattle's Broadview Green Grid. The offset design of the Pinehurst project treats runoff from 49 acres, while Broadview, designed with a curvilinear street, treats runoff from 32 acres, for the same cost of $4.6 million.

Carefully establish level of service. Consider impacts on the environment, feasibility, economic and social impacts, benefits and costs, and geographic needs, while establishing the level of service needed for a street.

OUTCOMES

Reduced runoff stormwater volume and peak flows during inundation events, and mitigated spot flooding within the project area.

Residents in the area have expressed appreciation for the aesthetic qualities of streets with natural drainage systems. The streets are favored places for neighborhood residents to take walks and are viewed as open space.

Water quality has been enhanced by a lower volume of stormwater leaving the site, and by biofiltration for excess stormwater that drains from the project area.

Pinehurst Green Grid, **SEATTLE, WA**

Commercial Shared Street

Shared streets prioritize walking over all other movement, while allowing motor vehicle access at extremely low speeds. Many narrow or crowded downtown streets operate informally as shared streets during rush hour or at lunchtime, but are not regulated as such. Commercial shared street environments are most suitable in places where pedestrian activity is high and vehicle volumes are either low or discouraged.

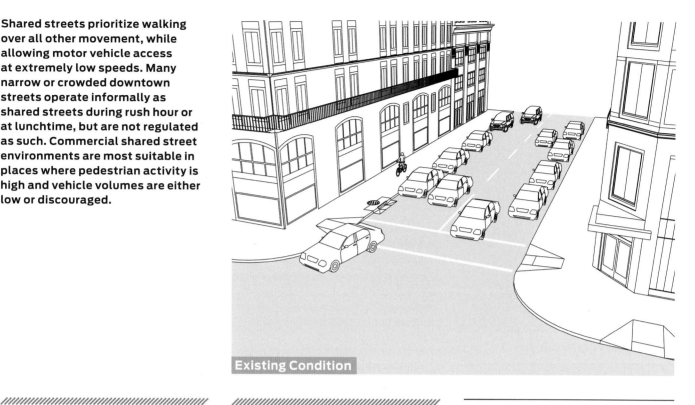

Existing Condition

EXISTING CONDITIONS

The illustrated condition serves underutilized commercial destinations. Sidewalks are narrow, often blocked, and the street is alternately crowded or deserted depending on the time of day.

On some wide downtown streets, speeding and cut-through traffic are an issue while on especially narrow streets, loading trucks block motor vehicle access entirely.

A flat street grade results in frequent standing water on the street, especially at low points such as curb ramps, gutters, and corner apexes, rendering the street inaccessible on a periodic basis.

RECOMMENDATIONS

Shared streets prioritize public space while providing safe access to all modes at very low speeds, with priority for people walking. Critically, these designs maintain access for loading and deliveries, and accommodate other motor vehicles. Design cues and local activity slow or divert traffic.

A curbless pedestrian and vehicle space with textured or permeable pavement reinforces the pedestrian-priority operation of the street. Use materials and vertical elements to delineate a shifting path of travel or narrow carriageway to encourage low speeds. Special pavements, especially permeable interlocking concrete pavers, may be subject to additional maintenance costs and should be selected based on regional climate and long-term durability. Selection of snowplow-compatible materials is recommended for colder climates.

1 Valley gutters or trench drains direct runoff to bioretention planters, and may be designed as detectable and high-contrast edges to delineate the shared roadway from the exclusive pedestrian paths.

2 If permeable pavement is used across the full roadway, design street grade and cross slope to channel water that does not infiltrate through the permeable pavement to flow to an approved discharge point.

3 Street furniture, including benches, planters, street lights, sculptures, trees, bicycle parking, and bollards if necessary, should be placed along the edge of a pedestrian-only space, subtly delineating the traveled way from the pedestrian-exclusive area. This edge should be designed carefully to be navigable by people who are blind or have low vision.

Local or state regulations may need to be enacted to enable shared street operations and enforcement, but designs should be self-enforcing on any shared street.

Commercial shared streets can be designed for a variety of widths, but the shared space itself should be kept narrow to limit traffic speeds.

On narrow shared streets and alleys, the entire surface is shared, and serves as the accessible pedestrian path, with a strict maximum speed of 5–10 mph.

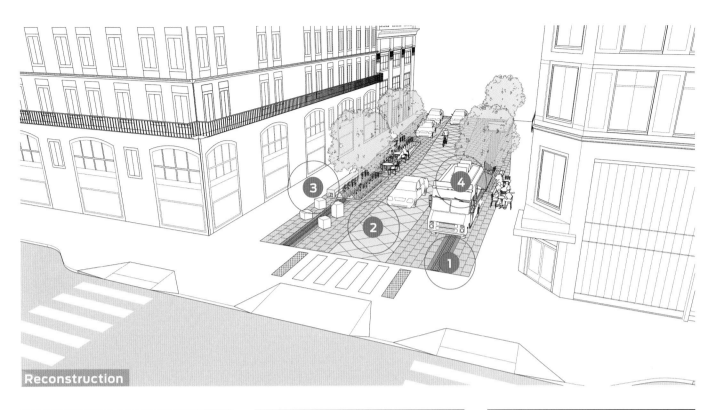

Reconstruction

On wider streets, consider providing a pedestrian-only clear path in place of a conventional sidewalk. Bollards, paving materials, and street furniture help to define parking spaces and delineate pedestrian and vehicular space. Staggered blocks of landscaping, bioretention facilities, or angled or perpendicular parking can be used to create a chicane effect. Special care should be taken to help visually impaired users distinguish zones that may be traversed by vehicles.

Bioretention facilities may need to be lined to prevent groundwater migration into adjacent structures and underground utility trenches. Review the condition of adjacent structures such as basements and utility corridors during the survey and planning process to determine seepage risk.

4 Commercial shared streets should be accessible to single-unit trucks making deliveries. Where commercial alleys are non-existent, it may be advantageous to design a shared street to accommodate large trucks as a control vehicle (not a design vehicle), though significant changes to the design should be avoided. Designated loading zones may be defined through differences in pavement pattern or use of striping and signage.

Provide tactile warning strips at the entrance to all shared spaces. Tactile warning strips must be placed where the accessible path of travel intersects with the motor vehicle travel path, as shown in the illustration above. On true single-surface streets without an exclusive pedestrian zone, warning strips should span the entire intersection crossing.

Use parking management strategies to ensure that parking activities and occupancy are suited to the commercial activity serving the street. Provide dedicated space or time of day for loading and deliveries, and site green infrastructure to allow up to a single unit delivery vehicle to access and load at commercial locations.

Partner with adjacent businesses or a business improvement district for programming and maintenance.

Potential GSI Features

Primary Roadway

» Permeable Pavement   ●

Planting Zone

» Bioretention Swale   ○

» Bioretention Planter   ●

» Tree Well or Trench   ○

Pedestrian Side Path

» Permeable Pavement   ○

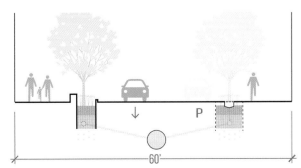

Case Study: Argyle Shared Street

Location: Chicago, IL

Street Context: Commercial Shared Street

Project Length: 0.26 miles

Right-of-Way Width: 66 feet

Participating Agencies: Chicago Department of Transportation, Chicago Department of Water Management

Timeline: 2012–2015

Cost: $4.8 million

GOALS:

Safety: Decrease traffic speeds and increase safety for pedestrians on this heavily traversed (by foot) corridor.

Placemaking: Create a strong sense of place with innovative streetscape design, providing a gathering place for the nearby community and supporting retailers in the project area.

Stormwater Management: Capture and infiltrate the majority of stormwater along the street.

Argyle Shared Street, **CHICAGO, IL**

OVERVIEW

The Argyle Shared Street project is Chicago's first shared street, designed to increase safety, provide the community with a plaza-like environment, and increase the attractiveness of the area for local businesses.

With plans for a full curb-to-curb reconstruction of the street, the city integrated extensive stormwater management features into the street redesign.

DESIGN DETAILS

A full reconstruction of the street raised the level of the roadway and eliminated curbs, creating a plaza-like effect, making the street fully ADA accessible, and providing for flexible street programming spaces.

35,000 square feet of permeable pavers delineate pedestrian, parking, and shared street space, visually narrow the street, and reduce the impermeability of the street.

Detectable edges delineate the vehicle and pedestrain spaces, **CHICAGO, IL**

Infiltration planters provide landscaping and soak up rainwater during storm events. On Argyle Street, 12 infiltration planters were built around existing catch basins, which were raised during construction. Eight additional infiltration planters are sited on side streets. Trees and plantings along the corridor provide shade and visual appeal, and capture and filter rainwater.

Argyle Shared Street, **CHICAGO, IL**

Three open-bottom catch basins located flush at alley entrances capture excess stormwater runoff in extreme rain events.

Chicanes and planters shift driving lanes slightly to the left and right, enhancing traffic-calming features of the street.

Sensors provide real-time information about the water management capabilities of the project. This data will help inform the design of future projects.

KEYS TO SUCCESS

Communicate the economic benefits of the project. Businesses along Argyle Street have been supportive of the project, in part due to clearly communicated benefits for them, including a more attractive streetscape, opportunities for seating and public events, and a broad recognition that most customers arrive on foot or by transit.

Foster an appropriate regulatory environment. An ordinance passed by Chicago's city council designated the project area as a "shared street," a new legal designation that allows for people on foot to cross anywhere on the street, not just at intersections. A separate ordinance permitted CDOT to set the street's speed limit at 20 mph.

Make sure all street users are well-served. CDOT worked with Lighthouse for the Blind and the Chicago Mayor's Office for People with Disabilities to ensure that the Argyle Shared Street is fully accessible to the blind, with extensive tactile warnings integrated into the design.

LESSONS LEARNED

Anticipate a learning curve. Drivers experienced confusion driving on, and especially parking on, the shared street in the first few weeks that it was open.

Perform outreach during and after implementation. Flyers, newsletters, an instructional video, and door-to-door visits were crucial to improving navigation and parking in the first weeks of the shared street's implementation.

OUTCOMES

The street redesign achieved a 31% reduction in impervious surface area.

During a typical storm event, 89% of stormwater is diverted from the gray infrastructure system.

Retailers reported an improved shopping and dining environment, with a more attractive streetscape and outdoor cafe seating.

Argyle Shared Street, **CHICAGO, IL**

Residential Shared Street

Many cities have primarily residential or other low-intensity streets where sidewalks and green infrastructure are either substandard or non-existent. These streets operate as de facto shared spaces, with people driving, bicycling, and walking in the roadway. Flooding and crumbling street surfaces are common.

These streets can often be redesigned as shared spaces, with a dramatically improved environment for walking, bicycling, and playing while accommodating service, delivery, and very local motor vehicle access.

Streets that are to operate as shared streets must be designed explicitly to promote safe, extremely low vehicle speeds.

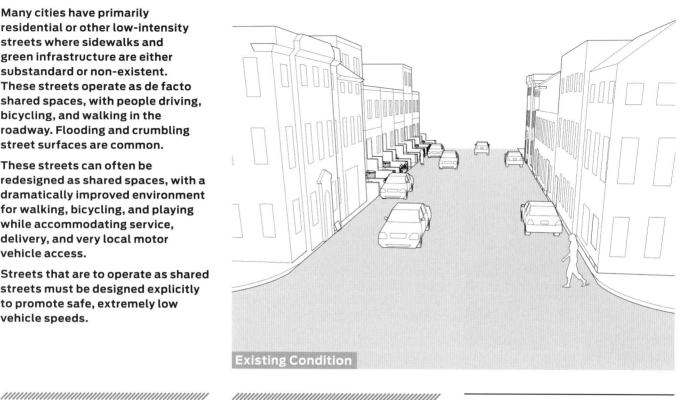

Existing Condition

EXISTING CONDITIONS

The roadbed accommodates low-volume local traffic. Parking occurs informally along the street and there may be unimproved drainage trenches.

Pedestrians use the entire road space for walking but are not provided any formal walking space or protection from through traffic. If any operational friction occurs (such as two vehicles passing, or a moving vehicle passing a parked vehicle), pedestrians are consigned to the edge of the street, in very uncomfortable walking conditions.

During large storm events, runoff may flood in the informal drainage trenches, gathering debris and pollutants and potentially pooling on the roadway.

RECOMMENDATIONS

1 Textured or permeable pavements that are flush with the curb reinforce the pedestrian-priority nature of the street. Special pavements, especially permeable interlocking concrete pavers, may be subject to additional maintenance costs and should be selected based on regional climate and long-term durability. Material selection should consider compatibility with winter maintenance, such as plowing, where applicable. If permeable pavements are used in the corridor, site facilities to minimize the amount of run-on from impervious areas.

2 In the illustrated example, trench drains collect runoff and direct into bioretention planters, while also providing a detectable alert between the shared roadbed and parking/pedestrian access lane. Shallow gutter pans may also be appropriate for directing runoff, depending on volume.

3 Grade the street to provide accessible, level walking surface while directing stormwater flow to green or gray stormwater infrastructure. The cross slope must be at least 1% to drain runoff, but cannot exceed 2% to provide an accessible pedestrian surface.

If permeable pavement is used across the full roadway, design the street grade and cross slope to channel water that does not infiltrate down the street to an approved discharge point.

Low fencing or slotted curbs may be sited around the entire edge of bioretention facilities to prevent incursion by pedestrians or vehicles. Bioretention facilities with vertical walls should be shallow in tighter geometries to reduce tripping and injury risk and allow for ease of maintenance.

4 Gateway treatments that slow or restrict traffic flow are critical to enforcing low-speed motor vehicle operation and provide safe and comfortable walking and bicycling conditions. Pinchpoints, speed humps, and raised crosswalks or speed tables are effective at enforcing motor vehicle speed, but must be designed to align with water flow patterns.

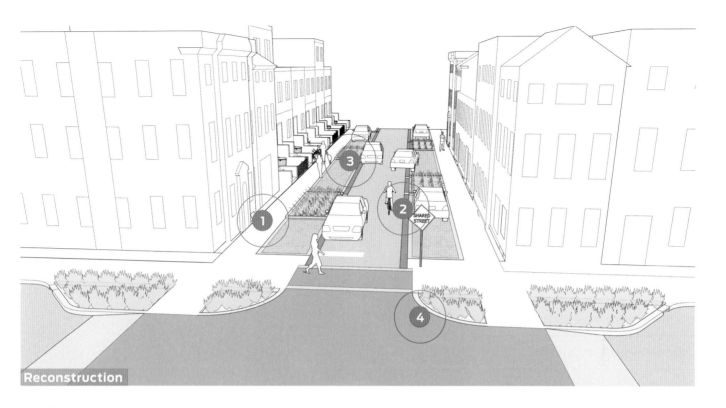

Reconstruction

Street furniture, including bollards, benches, planters, and bicycle parking, can help define a shared space, subtly delineating the traveled way from the pedestrian-only space. Bioretention can integrate seating or artistic elements to further define edging, protect stormwater facilities from incursion, and enhance the sense of place.

Bioretention facilities may need to be lined, especially when in close proximities to basements and structures. Implementation may require either relocating or installing a sleeve around service utilities near bioretention facilities.

In some cases, parking may be permitted directly adjacent to properties in a residential environment. Ensure that pedestrians are provided clear accessible paths along the entire surface of the street and to destinations at all times of the day.

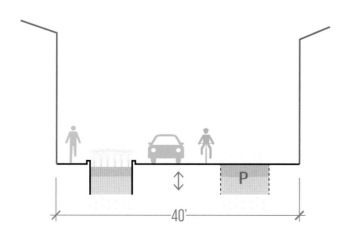

Potential GSI Features

Shared Roadway

» Permeable Pavement   ●

Planting Zone

» Bioretention Swale   ●

» Bioretention Planter   ●

» Tree Well or Trench   ○

Parking Lane

» Permeable Pavement   ●

Case Study: The Street Edge Alternatives (SEA) Street Pilot

Location: Seattle, WA

Street Context: Residential

Project Length: 660 feet

Contributing Drainage Area: 2.3 acres

Right-of-Way Width: 100 feet

Participating Agencies: Seattle Public Utilities, Seattle Department of Transportation

Timeline: Completed Spring 2001

Cost: $850,000

GOALS

Stormwater management: Reduce the two-year, 24-hour storm event (1.68-inches) peak runoff rate and volume to pre-developed conditions.

Environmental preservation: Reduce the flooding and pollution impacts of stormwater runoff on streams inhabited by salmon.

2nd Avenue NW, **SEATTLE, WA**

OVERVIEW

Seattle Public Utilities (SPU) fully reconstructed the street and drainage system on two blocks of 2nd Avenue NW to restore pre-development drainage that mimics the natural landscape. The project, located in the Piper's Creek watershed, aimed to mitigate the high flows and poor water quality that conventional gray infrastructure would discharge to Piper's Creek, a waterway that drains directly into Puget Sound, the second-largest estuary in the country.

The Street Edge Alternatives (SEA) Streets project introduced bioretention along a typical curbless neighborhood street with informal drainage infrastructure while simultaneously calming traffic along the street.

SPU worked collaboratively with residents to develop the final design, which successfully incorporated a number of green infrastructure techniques and provided a functional and livable community block.

DESIGN DETAILS

The project area's 25-foot-wide impervious road was redesigned into a 14-foot-wide curvilinear street, calming traffic and providing space for bioretention cells on both sides of the street.

The bioretention cells collectively achieve a detention volume of 2,500 cubic feet, and also allow for significant infiltration, which results in almost no stormwater runoff coming off the project block. The swales include three flow control structures with a 0.5-inch orifice at the outlet to reduce maintenance concerns; all flow control structures also include sumps to reduce the potential for clogging of the orifices. Some cells include an impermeable liner to prevent seepage into neighboring residences.

"Flat curbs" on both sides of the street provide additional space for emergency vehicle access without encouraging cut-through traffic on the residential street.

2nd Avenue NW, **SEATTLE, WA**

100 evergreen trees and 1,100 shrubs beautify the street with native flora and provide critical natural capacity to capture and filter excess stormwater.

2nd Avenue NW, **SEATTLE, WA**

LESSONS LEARNED

Rethink traditional approaches. Systems mimicking the natural environment can be as effective, or more effective, than a traditional curb-and-gutter system, and should especially be considered in less-developed areas installing new infrastructure.

Plan for startup costs. Initial outreach and coordination increased project costs, but laying this groundwork means that future projects will cost less than traditional street improvements.

Work with residents. Right-of-way needed to be reframed for residents who believed space was permanently earmarked for parking private vehicles, and who often landscaped the public area.

KEYS TO SUCCESS

Collaboration. Approach the project with a comprehensive, interdepartmental project team.

Community involvement. Involve neighborhood residents in all phases of design and construction, provide clear graphics and explanations, and build constructive partnerships.

OUTCOMES

Compared with a conventional curb-and-gutter system, the project reduced the volume of stormwater runoff by 99%.

Impervious surfaces were reduced by 11% compared with a conventional street.

While originally anticipated to attenuate up to 0.75 inches of rain, monitoring has shown the ability to attenuate substantially larger amounts.

Green Alley

Urban alleys, though often ignored or considered dirty or unsafe, can be designed to play an integral role in street networks, providing service access and recapturing space for the public realm.

Integrating green stormwater infrastructure into alleys transforms negative spaces into community assets that also serve mobility functions, improving the ease of access for service vehicles and freight and dramatically upgrading pedestrian and bicycle accessibility.

Existing Condition

EXISTING CONDITIONS

The majority of alleyways have low traffic and infrequent repaving cycles, resulting in back roads with potholes and puddling that are uninviting, unattractive, and almost always inaccessible.

Whether in a downtown or residential context, alleys often serve freight loading and deliveries, trash collection, or other large vehicle access that would otherwise use main street curb space. During the short periods where large vehicles are present, the alley becomes inaccessible to pedestrians using mobility devices who have difficulty crossing the curb.

Alleys may contain overhead power or underground utilities, which require bioretention facilities to be set back to maintain access for utility companies.

RECOMMENDATIONS

The design of green alleys should strive to balance their necessary utilitarian features with their placemaking potential. Green alleys use sustainable materials, pervious pavements, and effective drainage to create an inviting public space for people to walk, play, and interact.

1 Construct green alleys with permeable pavements such as pervious concrete or permeable interlocking concrete pavers with high reflectivity to reduce heat island effects and provide stormwater treatment. Avoid locating permeable pavements in areas where waste or recycling containers may be picked up to avoid having debris deposited onto the pervious surface. In areas with closed depressions that result in puddling, assuming the underlying soils are feasible for infiltration, permeable pavements could be used without having to install storm collection structures and an extension of the storm drain main.

2 Bioretention facilities can bring greenscape to the alley and create gathering spaces adjacent to the planters. This may require regrading the alley to direct flow that is usually down the center of the alley to a channel along the side and into the bioretention planter.

Bioretention facilities may need to be lined to prevent groundwater migration into adjacent structures and underground utility trenches. Review the condition of adjacent structures, such as presence of basements and utility corridors, during the survey and planning process.

3 Intersections between alleys and sidewalks have intentionally short sightlines, and should be designed to minimize the speed of motor vehicles (if permitted). Raise the intersection or the entire alley to the sidewalk grade and optionally use textured pavement to further reduce speed. Motor vehicle traffic in the alley approaching the sidewalk is typically controlled with a STOP sign (MUTCD R1–1).

Reconstruction

Green alley, **LOS ANGELES, CA**

Where permeable pavement or bioretention is not suitable, direct the alley drainage to a bioretention facility located in the planting strip on the downstream street.

To foster a safe, inviting street environment, green alleys should have human-scale lighting and clear sightlines to the street. Use light fixtures that focus their illumination toward the ground and minimize light pollution.

Alley greening and bioretention maintenance may be initiated and carried out by residents or neighborhood associations, or coordinated with city maintenance schedules.

Alleys provide direct property access and eliminate the need for driveways along main roads where people are walking and biking. Consider the use of alleys in all new developments and renovations to existing properties.

Green alleys have different maintenance responsibilities than conventional alleys. Use of textured pavements and other materials may be challenging to existing street sweepers and snowplows. Similar to shared streets, alleys may benefit from the application of snowplow-compatible materials and provisions for maintenance equipment access. Partnering with local businesses that are responsible for clearing any existing sidewalk of snow is often the best way to ensure prompt first-round snow clearance.

Freight vehicles may use green alleys for loading and unloading, which reduces double-parking on neighborhood streets. Low mountable or flush curbs provide pedestrians access around large vehicles.

Vehicle traffic can be restricted during non-delivery hours for outdoor seating or other uses. Where vehicle access is permitted, alleys should be designed for very low-speed movement with adequate space for delivery trucks and other freight vehicles. In narrow green alleys, permanent street furniture is minimal, allowing easy access by people and vehicles to any door, storefront, or loading dock. On wider green alleys, bollards and other street furniture and designated loading zones may be desirable; furnishings should not block loading paths. Since freight is often conveyed on hand trucks or cargo bicycles, careful attention should be paid to the location of curbs, if applicable, and access paths from loading zones to building entrances.

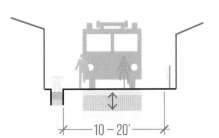

Potential GSI Features

Roadbed

» Permeable Pavement   ●

Property Line/Adjacent Properties

» Bioretention Planter   ●

» Permeable Pavement

10 – 20'

Case Study: *Johnson & Passenger Streets, Chattanooga*

Location: Chattanooga, TN

Street Context: Commercial Green Alley

Project Length: 400 feet

Right-of-Way Width: 47 feet

Participating Agencies: Public-Private Partnership between Chattanooga Department of Public Works and Crash Pad Hospitality

Timeline: 2012–2014

Cost: $350,000

GOALS

Stormwater management: Capture the first inch of runoff during storm events and decrease the peak runoff rate during extreme rain events.

Flooding control: Eliminate nuisance flooding events in the project area.

Economic cost: Provide a more cost-effective stormwater solution than a traditional gray infrastructure treatment.

Johnson Street, **CHATTANOOGA, TN**

OVERVIEW

The Tennessee Department of Environment and Conservation has mandated that Chattanooga refurbish its sewer and stormwater infrastructure due to combined sewer overflow (CSO) events into the Tennessee River, a source of drinking water for communities across the Southeast U.S. As a result, Chattanooga requires that all new developments within the combined sewer system reduce or detain stormwater runoff during rain events.

Crash Pad Hospitality, owners of a nearby popular hostel, started planning for a restaurant (The Flying Squirrel, now open). Early on, it became apparent that Johnson Street would have to be refurbished to allow the restaurant to open, as it had become a dilapidated street with significant drainage issues, causing health and safety issues for the restaurant, the hostel, and the surrounding community.

A traditional street reconstruction with new gray stormwater infrastructure was estimated to cost $327,000, and would not have included any water quality improvements or peak flow reduction during rain events, required under city rules for new development. The City of Chattanooga and Crash Pad Hospitality instead agreed to a public-private partnership to transform Johnson Street into a bicycle and pedestrian friendly green street with significant green stormwater infrastructure.

The design accommodates runoff from all nearby properties; adjacent businesses can "buy-in" to the system to avoid the need to site stormwater management practices on high-value real estate.

DESIGN DETAILS

Over 14,000 square feet of permeable brick pavers, paid for by the private partner, capture stormwater from the street and adjacent land.

Three feet of gravel under the street surface captures stormwater and convey it to the underlying subgrade soil at a slow rate. Excess stormwater from extreme rain events is conveyed through an 8-inch underdrain system, substantially smaller than what would be required in a traditional gray infrastructure approach.

Taking advantage of the need to fully reconstruct the street, Johnson Street was converted into a shared street, comfortable for pedestrians and bicyclists, with a design that naturally slows down motor vehicle traffic.

KEYS TO SUCCESS

Work closely with partners. A strong working relationship between the City of Chattanooga and the private partner, Crash Pad Hospitality, allowed for an ambitious, innovative design.

Use opportunities to accomplish multiple street goals. The final design is a strong asset both for the private developer as well as for the city, transforming a flood-prone street into a rain sponge that's equally comfortable for pedestrians and bicyclists, while maintaining (and improving) freight and loading access for adjacent businesses.

OUTCOMES

The project captures 11,320 cubic feet of stormwater, equivalent to 2.25 inches of runoff, per storm event.

The project is abating more stormwater than required by the new development, allowing future development along the corridor to buy into the project instead of building costly gray infrastructure.

Before reconstruction

After reconstruction

Industrial Street

Large industrial streets serve manufacturing, warehouses, and heavy commercial land uses. These streets have historically moved truck traffic and provided little or no accommodation for people walking and bicycling.

As economies shift from manufacturing to local producers and retail-wholesale mixed with hospitality and entertainment, many North American cities are seizing the opportunity to revitalize industrial streets. While they still operate as trucking routes, bike freight distribution is growing quickly, and the street is becoming a destination in its own right.

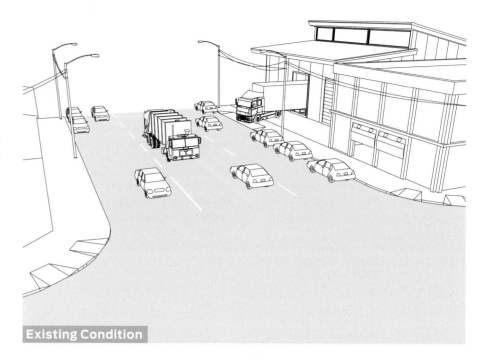

Existing Condition

EXISTING CONDITIONS

The street relies entirely on the gray stormwater system, and there are few if any trees, resulting in uncomfortably hot and cold conditions. The roadbed is designed primarily to serve heavy vehicle traffic, with little or no accommodation for people biking, walking, or taking transit. Few people use the street voluntarily.

Minimal pedestrian facilities are provided, and existing ones are often blocked by trucks. Key bus routes serve the street and arriving workers, but stops have low ridership and no shelters. Riders face inhospitable waiting and crossing conditions.

Stormwater runoff entering the gray infrastructure system has large amounts of sediment, debris, and pollutants from the roadway that are either treated off-site or directed to waterways during overflow events.

RECOMMENDATIONS

1 The street has been converted from two lanes in each direction to one, with a left turn pocket where needed. Seize opportunities to align provision of safer bicycling and walking conditions with improved stormwater infrastructure.

Use green expressions, including bioretention facilities and street trees, to improve the experience of walking, bicycling, and riding transit. Integrate green infrastructure into transit stops and the planting zone to help capture and dissipate water and air pollution from the street.

2 Shorten crossing distances and tighten curb radii to improve pedestrian safety. Where large vehicles are expected to make turns, mountable corner aprons or concrete "pillows" can allow large vehicles to make turns while discouraging car drivers from making high-speed turns.

3 Adding bikeways can reveal new space within the cross-section to incorporate green infrastructure, such as bioretention cells in the bikeway buffer zone (if adequate width is available, typically at least 5 feet). Bikeways are also appropriate for permeable paving treatments since people biking incur less wear-and-tear on the surface than do vehicles. Utilize pervious concrete or porous asphalt to ensure the surface is compatible and comfortable for bicyclists.

If bikeways are configured against the curb, design with sufficient width to keep bicycles from operating in the curb or gutter. The preferred minimum width for one-way curbside bikeways to provide comfortable riding is 6 feet.

Industrial streets with high truck traffic generate large sediment and debris loads depositing into bioretention facilities, requiring large presettling zones at inflow points.

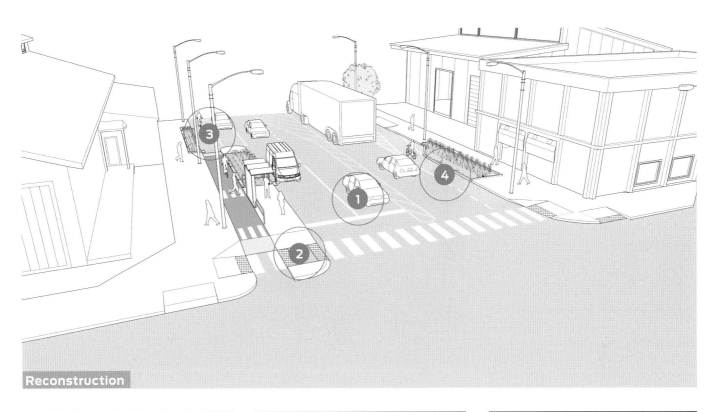

Reconstruction

Test existing soils for pollutants early in the project design phase, as certain areas may have contaminants that may need to be remediated before infiltrating stormwater facilities are installed or pollutants may be transferred through the air onto the roadway from adjacent use. As industrial sites redevelop, opportunities emerge to capture and mitigate certain pollutants.

The soil media in bioretention facilities in industrial areas may be multi-layered or a different mix than that used on residential yield streets, depending upon the pollutants that are intended to be treated (both from the air and street runoff).

Bioretention facilities may need to be lined to prevent existing contaminants in the soil from becoming mobile and infiltrating further into the ground.

Target installation of high-volume bioretention facilities to store and treat larger amounts of runoff.

Assess the street network for opportunities to convey water to remote sites where more water quality treatment can be implemented in large swales.

④ Parking lanes and bikeways can often accommodate permeable surface. Limit the amount of stormwater run-on from the travel lanes onto permeable pavement parking lanes and bikeways, since higher sediment and pollutant loads from industrial travel lanes will incur more frequent maintenance requirements.

Potential GSI Features

Median

» Bioretention Swale

Curb Extension

» Bioretention Planter

Planting Zone

» Bioretention Planter

» Tree Trench

Permeable Pavement

» Parking or Bike Lane

» Primary Roadway

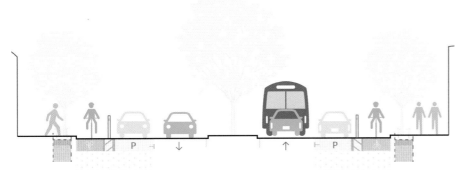

Case Study: Cermak & Blue Island

Location: Chicago, IL

Street Context: Industrial

Project Length: 1.4 miles

Right-of-Way Width: 100 feet

Participating Agencies: Chicago Department of Transportation; Chicago Department of Planning & Development; Metropolitan Water Reclamation District

Timeline: 2009–2012

Cost: $14 million

GOALS

Stormwater management: Divert 80% of stormwater away from graywater infrastructure.

Demonstrate possibilities: Serve as a scalable model for future sustainable streets projects in Chicago.

OVERVIEW

The Cermak/Blue Island Sustainable Street project transformed a 1.4-mile stretch of faltering industrial route into one of the greenest streets in the country, while piloting a process for future green street transformations that maximize the environmental and mobility performance of city rights-of-way.

Using a holistic systems approach, the project team implemented a street that embodied sustainability from design to materials selection and production to construction to performance.

Street trees, bioretention planters, and permeable pavements on Blue Island Avenue, **CHICAGO, IL**

The reconstruction of a 1.4-mile section of Blue Island Avenue and Cermak Road in Chicago's Pilsen neighborhood is CDOT's first sustainable street project, a pilot that sought to demonstrate how to design, source, and construct an environmentally productive, inviting, and safe urban street on an especially challenging corridor. The project street runs through a heavily industrial area adjacent to an active freight railway and still serves a significant volume of heavy truck traffic.

The project team had to balance their desire to transform the section into right-of-way that collected and retained large amounts of water—thus limiting the likelihood of outfalls into local waterways—while still serving the mobility and economic needs of the corridor.

Designers applied a "belt-and-suspenders" approach to the project: rather than simply installing standard swales or rain gardens, the team conducted a thorough investigation and designed to maximize water infiltration. Each piece of the design sought to contribute to the overarching goal of the project: to make Cermak/Blue Island the "greenest street in America."

DESIGN DETAILS

Because the south side of the Cermak segment is adjacent to an active railway that precluded full sidewalk installation, the project seized the chance to implement continuous swales to capture large amounts of runoff.

Large infiltration planters were implemented on the north side of Cermak and along Blue Island in the sidewalk furnishing zone. Blue Island Avenue included changes to the curb line to accommodate bulb-out planters, narrowing the roadway and enhancing the pedestrian realm.

Infiltration planters were sited over catch basins, with two inlets surrounding each basin to direct as much runoff as possible into planters before reaching the graywater system. Bioretention cells included both a primary infiltration gallery directly beneath the cell and a secondary gallery under the sidewalks. This aggressive water storage capacity allows more than 80% of stormwater to be captured and stored, despite relatively poor infiltration in native soil.

The parking and bike lanes were striped over pervious pavers, allowing greater water capture and infiltration from the 80-foot-wide roadway. The roadway itself is 40% composed of high albedo (highly reflective) pavement surfaces to reduce the street's urban heat island effect.

Additionally, the roadway incorporated a photocatalyic microsurface, which actively reduces nitrous oxide pollution over the street, improving air quality.

Bikeways and parking lanes are surfaced with interlocking permeable pavers.

The project also introduced improved walking and bicycling conditions (including wider sidewalks, curb extensions, refuge islands, and new bike lanes), new public plazas and seating, and enhanced transit stops with high quality shelters. Curb extensions along the full project extent and the refuge island in front of Benito Juarez Community Academy, a high school, have calmed and improved traffic conditions while improving mobility options.

Finally, the reconstruction created two new public plazas, which include large rain gardens to capture and treat water, along with valuable public space and seating to activate a formerly underutilized industrial street.

Bioretention planters are integrated with public realm and walking environment.

LESSONS LEARNED

Apply a systems approach. Getting to a high performance project requires doing the necessary engineering to fully integrate strategies. All bioretention and street design elements are interlocking and contribute to the total impact.

Demonstrate the benefits of investment. Stakeholders supported the larger project cost because they understood the benefits of a more productive street over the long term.

Make internal coordination a habit. By holding weekly intradepartmental meetings, different sections of CDOT maintain a clear wide-angle picture of projects and involvement. Clear communication enables synergy and reduces duplication or conflict.

Minimize utility impacts. CDOT has a policy that once a street is reconstructed, there is a moratorium on street cuts in perpetuity, requiring utilities and related agencies to coordinate infrastructure projects during the full reconstruction. Additionally, the City was able to design infiltration planters so that utility lines are able to continue through, allowing more efficient use of sought-after street width.

Solar and wind energy collection.

OUTCOMES

The project is exceeding its original goal of infiltrating 80% of stormwater runoff along the corridor.

New partnerships were enacted to maintain and monitor the street: the MWRD is partnering with the USGS to collect volume and water quality data at four wells, and CDOT partnered with the local business district to hand off maintenance after a two-year establishment period.

Retrofits, including bioretention, new LED lighting, and solar and wind-powered bus shelters, reduced the energy use of the street by 42%.

The project successfully leveraged techniques to close waste loops: recycled materials make up 23% of the reconstructed roadway, and the construction team recycled 60% of construction waste.

The City effectively utilized local and regional businesses and supply chains to reduce embodied implementation impacts and spur local growth: 76% of raw materials were sourced and manufactured within 500 miles of the project site, and 23% were sourced within 200 miles.

The project was 21% less expensive than Chicago's average reconstruction of similar scale during the year it was completed, even with the multitude of innovative design features.

Stormwater Greenway

Cities around the world are restoring urban streams that had previously been decked or canalized, often with a roadway on top.

Streets-to-streams projects are transformative, presenting an opportunity to implement high-performance water quality management practices, while creating inviting and active public spaces. These spaces can become destinations in themselves, giving people in cities access to a new kind of waterfront.

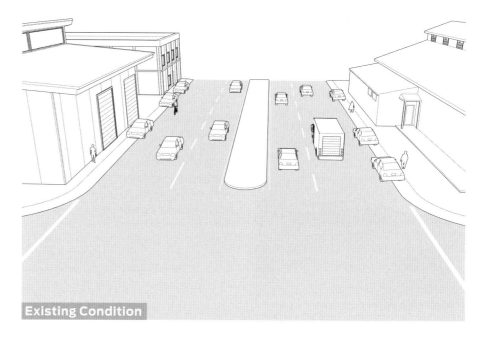

Existing Condition

EXISTING CONDITIONS

The street, initially designed as a primary corridor aligned with a stream, has been decked over or channelized, often into a large subsurface pipe or aqueduct.

Substandard or missing sidewalks and no bike facilities create an inhospitable and dangerous environment for all users. Vehicles routinely park on poorly defined curbs. Speeding is a consistent issue.

Poor or depleted pavement quality along the flow line contributes to frequent ponding, further degrading pedestrian accessibility and pollutant accumulation.

The complete absence of tree canopy contributes to a local heat island effect.

Stormwater is directed entirely to the graywater system, resulting in regular overflows that send untreated runoff into the regional watershed, degrading water quality and putting greater strain on off-site treatment facilities.

RECOMMENDATIONS

1 Restore the natural drainage swale and daylight the stormwater flows to fully express water in the streetscape, provide street beautification, and create usable public space.

Large stormwater facilities create space for innovative or high-performing stormwater facilities, including water quality treatment, fish and amphibian habitat restoration, and recreation space. Design facilities for the appropriate context, including local flora, drought risk, and community needs. Continue streams through crossings and connect with regional waterways.

2 While upstream flow is directed into the median stormwater stream, local street runoff is directed into curbside green infrastructure, including curb extension planters and pervious surface on the raised bikeway. Subsurface infrastructure can direct overflow into the median drainage swale.

3 Direct gutter flow along raised crossings to appropriately sized inlets to mitigate risk for pooling; cover inlets with metal lids to prevent wheels from entering planters.

4 Pervious pavement on the bikeway infiltrates sheet flow from the sidewalk and water that falls on the bikeway. Raised cycle tracks, with raised crossings, can provide comfortable bicycling conditions for all ages and abilities when properly designed (refer to the *Urban Bikeway Design Guide* for additional guidance) while performing stormwater management functions and creating a more active and inviting public space.

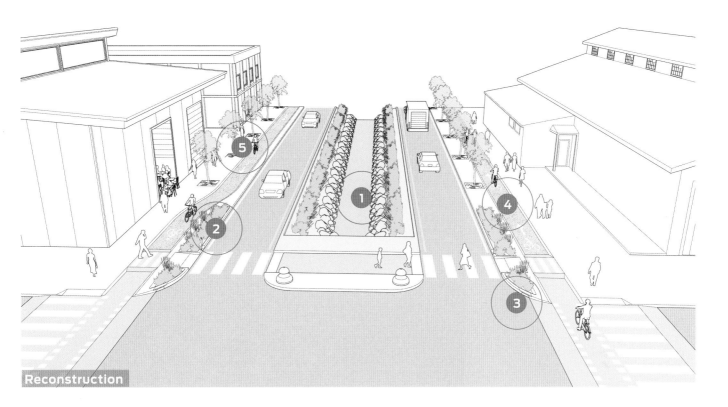

Reconstruction

5 Trees can be planted either between the bikeway and pedestrian path, or between the bikeway and street. Design tree pits or wells to mitigate risk of pedestrians or bicyclists entering accidentally, as well as risk of bicycle pedals striking raised curbs, lips, or short fencing. Tree grates can be used to protect soil against compaction.

Ensure trees planted in sidewalk/ planting zone tree wells have sufficient subgrade root space to reduce the likelihood of broken or launched sidewalks. Select trees so that branch height or width does not invade either the pedestrian path or bikeway.

Potential GSI Features

Median

» Bioretention Swale or Waterway ●

Travel Lanes / Roadbed

» Permeable Pavement ○

Curb Extension

» Bioretention Planter ●

Planting Zone

» Bioretention Planter ●

» Bioretention Swale ●

» Tree Well or Trench ●

Bikeway / Shared Use Path

» Permeable Pavement ●

96'

Case Study: Ed P. Reyes River Greenway

Location: Los Angeles, CA

Street Context: Stormwater Greenway

Project Area: 1.13 acres

Drainage Area: 135 acres

Right-of-Way Width: 100 feet

Participating Agencies: LA Bureau of Sanitation (LASAN), Watershed Protection Division

Timeline: Opened to public 2013

Cost: $3.4 million

GOALS

Stormwater management: Capture all dry weather runoff from the project area.

Improve water quality: Divert and filter polluted stormwater to mitigate the highest bacteria point source before contaminants reach the Los Angeles River.

Increase open space: In a neighborhood of Los Angeles starved for open space and greenery, the stormwater facility serves a function akin to a public park and provides public access.

Seating & public space, **LOS ANGELES, CA**

Ed P. Reyes River Greenway, **LOS ANGELES, CA**

OVERVIEW

The project was identified by a regional study, Cleaner Rivers Through Effective Stakeholder TMDL's, or CREST. The site is located on the north bank of the Los Angeles River in the Lincoln Heights community, a neighborhood underserved by open, green spaces. The project is situated in an undeveloped section of right-of-way at the end of Humboldt Avenue, a location that has long been problematic for the City of Los Angeles because it has become an illegal dumping ground and hosting various encampments.

An existing storm drain spanning the project site delivered runoff from a 135-acre tributary to an outfall into the Los Angeles River. Stormwater runoff from streets and highways, industrial facilities, multi-family residential structures, and commercial structures conveyed pollutants to a federally regulated water body, enforced by the US EPA.

Through landscape interventions, this former brownfield site and storm drain was daylighted and transformed into an ecological resource closer to a natural tributary. Much of the site has been opened to the surrounding community as a community facility with walking paths and a small pedestrian bridge, with opening hours every day.

DESIGN DETAILS

A hydrodynamic separator and a solar pump lift stormwater runoff from a wet well to a level spreader and spillway ("waterfall"). Water flows to a vegetated forebay, which provides year-round support for flora and fauna.

In wet-weather events, storm flows overtop a rock weir to infiltrate a sub-surface bio-infiltration gallery beneath 18" of vegetated soils. In extreme stormwater events, flows rise above grade to the standpipe and flow to the Los Angeles River.

Runoff is piped into the stormwater treatment facility, **LOS ANGELES, CA**

LESSONS LEARNED

Have a clear plan for public access, operations, and maintenance.

A security vulnerability initially limited public access to the site until a plan could be put in place for safe access. Monitoring, operational training, and ongoing coordination between stormwater utility crews has enabled public access to the space during daylight hours, and extends public access in summer months. As the lead agency for the city's stormwater compliance, LASAN is responsible for permit compliance and so oversees operations and maintenance of the facility.

OUTCOMES

The basin, storage, and filtration structures detain approximately 50,000 cubic feet of stormwater runoff.

The project achieved full compliance for dry weather bacteria flows from the tributary to the Los Angeles River by capturing 100% of flows and filtering and using them to support the ecosystem functions.

The project is serving as a much-needed open space resource for a disadvantaged community.

Multi-use trails and public space are integrated into stormwater facility design, **LOS ANGELES, CA**

Educational signs, **LOS ANGELES, CA**

Reclaimed Intersection

Complex or multi-legged intersections are common in non-grid street networks, or where two grids at different orientations come together. Multi-legged intersections can be reconfigured to improved access for people walking and bicycling, while capturing large amounts of surface area to gather and infiltrate runoff.

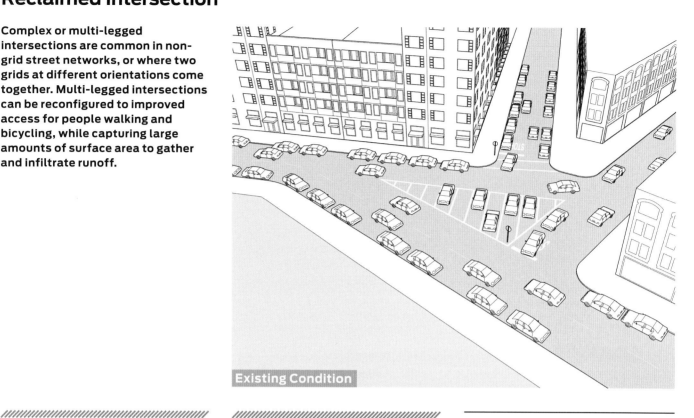

Existing Condition

EXISTING CONDITIONS

In the condition shown, two neighborhood streets come together at a larger collector street, resulting in difficult crossings for pedestrians who are forced to either detour or cross at an unmarked location.

People bicycling and driving encounter a confusing circulation pattern. Additionally, vehicles may park (legally or illegally) in the center island space, restricting sight-lines.

The footprint of the junction is a large swath of impermeable streetscape, with most runoff being directed past pedestrian crossings and increasing incidents of pooling.

RECOMMENDATIONS

Observe intersection movements and desire lines. Identify opportunities to capture excess impermeable space in the roadbed and organize traffic movements. Enable pedestrians and people bicycling to move along natural desire lines or intuitive paths through complex intersections.

Test traffic reconfigurations with tactical or interim treatments before making capital investments.

1 Reduce the number of conflict points between motor vehicles and other users. In the illustrated example, motor vehicles are routed through turns sequentially to simplify signalization and turning movements. Pedestrians are given more space to walk, and are provided crossing treatments at expected locations.

2 In the recaptured plaza space, stormwater can be directed and retained in a high-capacity graded facility, adding green space to a formerly imposing streetscape, and treating and infiltrating a large amount of water. The green infrastructure provides a large amount of edging which can incorporate seating and placemaking elements to engage users. Stationary activities can be encouraged by providing flexible, movable seating and tables in plaza areas. Newly reclaimed street space can also be used to site bike share stations.

Incorporate shade trees to provide canopy and encourage people to stay and use plazas and the public realm.

3 Tree wells or connected tree trenches can capture and infiltrate runoff while contributing to tree canopy and increasing walking comfort. Because tree wells have a smaller footprint, they may be included more easily where sidewalk width is limited or pedestrian volumes are moderate-to-high. Consider using low fencing or curbs to prevent accidental incursions by pedestrians, and site clear of crossings and curb ramps.

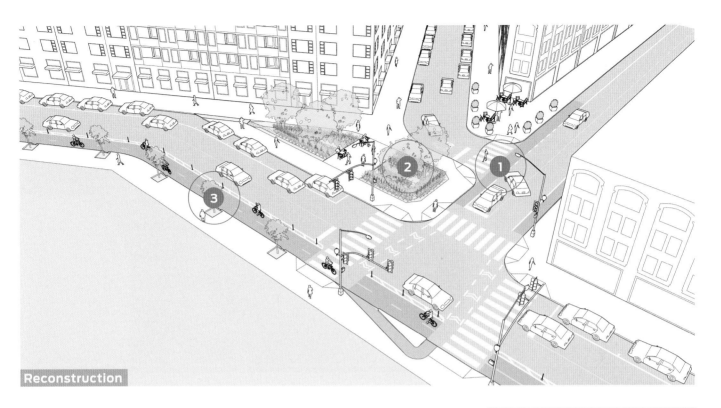

Reconstruction

Avoid siting drainage grates and catch basins at corner apexes or in places where pedestrians are likely to trip when stepping on or off the curb.

Potential GSI Features

Plaza / Curb Extension

» Bioretention Planter   ●

» Bioretention Swale   ●

» Tree Trench   ●

Planting Zone

» Bioretention Planter   ●

» Bioretention Swale   ●

» Tree Well or Trench   ●

Bike / Parking Lane

» Permeable Pavement   ●

NE 47th Avenue & NE Euclid Avenue, **PORTLAND, OR**

Case Study: Washington Lane & Stenton Avenue

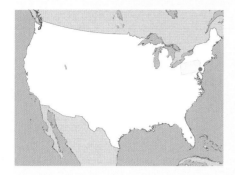

Location: Philadelphia, PA

Street Context: Reclaimed Intersection

Project Area: 0.2 acres

Drainage Area: 0.3 acres

Right-of-Way Width: 42 feet

Participating Agencies: Philadelphia Streets Department, Philadelphia Water Department

Timeline: Opened 2014

Cost: $420,000 ($300,000 for street reconstruction, $120,000 for green infrastructure)

GOALS

Stormwater management: Capture stormwater runoff from adjacent right-of-way.

Improve water quality: Divert and filter polluted stormwater before it reaches the Tookany-Tacony Frankford Watershed.

Promote traffic safety: Eliminate slip lanes, increase pedestrian amenities, and re-time signals.

Pilot green streets: Provide a case study for collaborative investments between departments.

Washington Lane & Stenton Avenue, **PHILADELPHIA, PA**

OVERVIEW

Stenton Avenue is a major thoroughfare travelling through Northwest Philadelphia. In 2013, the Streets Department began construction of corridor-wide signal re-timing and other safety investments. The Streets Department approached the Water Department early in the design phase about the opportunity to add green infrastructure at select intersections, pairing stormwater management with geometric changes to improve traffic safety.

The Water Department developed a rain garden to fit within the Streets Department's proposal to remove a slip lane at Stenton Avenue and East Washington Lane. Working with local community groups, the final project incorporated a small plaza area on this major intersection, bringing both green and pedestrian amenities to this corridor.

DESIGN DETAILS

The slip lane, which previously enabled right turns at speed from Stenton Avenue onto East Washington Lane, was closed and filled in with new rain gardens and sidewalk space. In addition to reducing vehicle speeds at the right turn with the tighter turn radius, the reconfiguration straightened the walking path for pedestrians crossing East Washington Lane and shortened the crossing distance. The slip lane closure also removed a conflict point between the northbound bike lane and merging motor vehicle traffic from Stenton Avenue, improving safe conditions for all users.

The three rain gardens, which are constructed with graded sides and a large infiltration area, store and clean stormwater runoff while improving the pedestrian environment. Each of the bioretention cells is 18 inches deep. A trench drain conveys water from the right-of-way to the rain gardens with stone storage below each garden, each separated by a different trench drain.

A pedestrian path was constructed bisecting the Stenton Avenue-facing rain garden from the two rain gardens along Washington Lane, providing a direct pathway from the storefronts at the corner to the intersection crossings. The path has a raised curb to provide a detectable edge for pedestrians, while the edges along the street sides are planted with low shrubs to provide a soft barrier between pedestrians and the bioretention space.

An overflow control structure manages excess stormwater at the connection to the sewer system.

LESSONS LEARNED

Interdepartmental communication promotes iterative green street deployment. Strong working relationships between departments promote green street deployment and support cost sharing, delivering projects with a triple-bottom-line value proposition.

OUTCOMES

The basin, storage, and filtration structures manage approximately 2,326 cubic feet of stormwater runoff.

After demonstrating successful implementation at this location, the Philadelphia Water and Streets Departments have leveraged this collaboration to repurpose right-of-way space in multiple other projects, including squaring a large intersection corner at Trenton Avenue & Norris Street (shown at right), and working alongside other partners to incorporate stormwater management systems into a high-quality bicycle and pedestrian facility on the Penn Street Trail (shown at far right).

Stormwater runoff is distributed among three bioretention areas for infiltration, Washington Lane & Stenton Avenue, **PHILADELPHIA, PA**

Trenton Avenue & Norris Street, **PHILADELPHIA, PA**

Penn Street Trail at Delaware Avenue & N Penn Street, before & after, **PHILADELPHIA, PA**

Yale Avenue North, SEATTLE, WA

4 Stormwater Elements

Newcomb Ave, SAN FRANCISCO, CA

Green Stormwater Elements

The green stormwater infrastructure toolbox includes a variety of design elements that must be selected, sized, and configured to meet the goals and context of the project site. Multiple green elements may be sited and combined within the street to realize the full potential to manage stormwater runoff, improve multi-modal mobility, enhance street aesthetics, and achieve the full performative value of living infrastructure.

Bioretention Planter

Bioretention planters are stormwater infiltration cells constructed with walled vertical sides, a flat bottom area, and a large surface capacity to capture, treat, and manage stormwater runoff from the street.

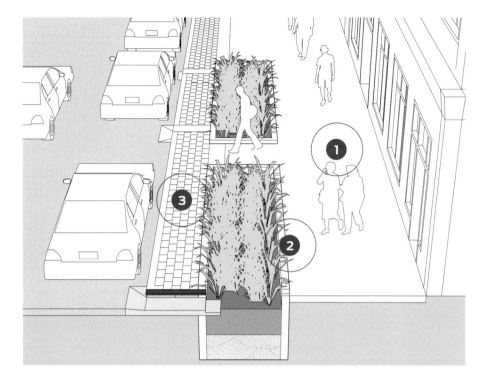

APPLICATION

Bioretention planters are cells with walled vertical sides. They maximize the bottom area of the cell and the volume of water that can be temporarily stored.

Planters can be implemented nearly anywhere in the right-of-way, including in the sidewalk furnishing zone, in medians and pedestrian boulevards, or along the property line.

Planters can be adapted to widely varying urban street contexts, with flexible depth, edge construction, and vegetation. Planters are effective where right-of-way width is constrained or multi-modal capacity is high.

BENEFITS

Bioretention planters offer greater capacity within the cross-section for stormwater detention and infiltration than bioretention swales.

Planters are highly adaptable to most urban contexts, and can be sized and modified easily to optimize infiltration rate in constrained spaces.

CONSIDERATIONS

The height of the vertical wall depends upon the design ponding depth, slope of the road, and bottom slope of the facility. Consider the need for operations and maintenance; crews will need to access the bottom of the facility. Also consider whether having a vertical drop adjacent to the sidewalk and roadway will affect general comfort for people using the street.

The vertical walls are commonly concrete (either prefabricated or cast in place) but other materials (e.g., rockeries) may be used.

Water may migrate laterally as it infiltrates into the media, and may be at risk of entering adjacent basements and structures. Increase wall depth or use liners to mitigate water migration.

New planters may need to be irrigated during an establishment period of 1–3 years; reclaimed water may be used to offset resource needs, but that requires additional subsurface infrastructure.

CRITICAL

When siting curbside planters, maintain pedestrian access and capacity. Bioretention cells must not encroach upon clear walking paths, and cannot impede designated accessible parking spaces or loading zones.[1]

Bioretention planters must be sized to handle the runoff load of the design tributary area (see Bioretention Cell Sizing on page 102). Multiple planters may be linked or sequenced to prevent overloading on a single planter.

Bioretention planters must drain within 24–72 hours (depending upon frequency of rainfall) after a storm event to prevent insect breeding and bacteria or algae formation. (See page 103 for additional discussion of drawdown time.)

RECOMMENDED

The planter bottom should typically be at least 4 feet wide to promote vegetation health. Narrower cells may be implemented in special contexts, such as protected bikeway buffer zones or constrained sidewalks, but must be designed with consideration for plant health, bioretention performance, and implementation cost.

The maximum ponding depth (see page 100) is typically not more than 6–9 inches, with 12 inches from sidewalk to soil in areas with high pedestrian activity to mitigate fall risks. Deeper bioretention cells require more robust barriers, such as fencing or railings.[2]

1 Where pedestrian activity is moderate or high, at least 8–12 feet of clear width should be dedicated to pedestrian movement between the edge of the planter and the building or property line. Where pedestrian activity is low, a 5- to 6-foot preferred minimum allows two people to pass comfortably.

2 Provide an edge that can be detected using a cane or other mobility device to prevent incursions, such as a 4-inch curb or low fencing (less than 24 inches tall).

Plants that grow to at least the height of the wall (preferably with some variable height) act as a visual barrier and discourage incursions.

Planter step-out zone, **PHILADELPHIA, PA**

3 Planters configured adjacent to on-street parking should provide a level step-out zone along the curb to accommodate vehicle entry and exit, and mitigate soil compaction and trampling, typically 36 inches wide from the curb face. Provide regular access paths between the curb and pedestrian through zone, at least 5 feet wide.

Shorter planters (20 feet or less in length) with regular access paths between the curb and sidewalk can enable narrower step-out width on constrained sidewalks; 12–18 inches of firm, compacted surface behind the curb face can accommodate vehicle entry and access. Provide designated parking and loading spaces with accessible paths and clearances for people using mobility devices on every block and at major destinations.

Planter with a 12-inch stone forebay behind the curb, **NEW YORK, NY**

OPTIONAL

To provide shade canopy and improve walking comfort, street trees can be integrated into walled stormwater planters, increasing transpiration and water volume managed. Ensure trees have adequate root space for health, and select appropriate species for local climate and cell ponding characteristics. Medium and large trees should be sited outside planters unless adequate clearance from walls and soil volume can be provided to allow for adequate root space (see page 84).

Seating, informational signs, and other urban design features may be utilized as barriers around deeper planters to improve pedestrian safety while also beautifying the streetscape.

Informational signage raises visibility and engages the public, **PORTLAND, OR**

TYPICAL DIMENSIONS (NOT ILLUSTRATED TO SCALE)

PARKING-ADJACENT PLANTER

18 - 36 IN. STEP-OUT · 12 – 18 IN MAX · 48 IN. PREFERRED MIN · 4 IN. · 6 – 9 IN. TYP · 12 IN. MAX · .5 - 2%

Planters adjacent to parking should provide a curbside step-out of at least 18 inches, though up to 3 feet may be preferred. Cross-slopes draining into the planter must be between 0.5–2%.

CURBSIDE TRAVEL LANE-ADJACENT PLANTER

12-24 IN · 6 IN. MIN · 2 IN. DROP · 12 - 18 IN MAX · 4 IN. · 6 – 9 IN. TYP · 12 IN. MAX · 6 - 12 FT. DESIRED

Planters adjacent to a curbside travel lane may not need additional offset from the curb (a 2-foot level area may simplify maintenance access). Low curb or fencing may be used to protect against entry.

Biofiltration Planter

Where infiltration cannot be accomplished due to contextual characteristics, native soils, or other constraints, walled planters can be designed with an impermeable base and supporting drainage infrastructure that collects water, filters runoff downward through soil media, and channels treated runoff through an underdrain (perforated) pipe. Biofiltration planters provide water quality treatment and reduce runoff volumes, and may be applied in more limited rights-of-way.

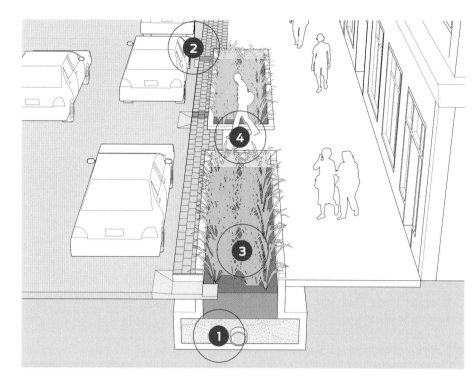

APPLICATION

In areas where infiltration of stormwater is not conducive—including constrained sites next to buildings or with setback limitations, where existing soil conditions limit infiltration, adjacent to steep slopes (>4%), or in areas with contaminated soils—non-infiltrating bioretention planters with a subsurface drain system may be utilized to still manage stormwater volume and treat water quality.

The bottom of the bioretention planter at subgrade may be made of concrete, or it may have a liner attached to all sides of the planter in order to deter water from infiltrating.

BENEFITS

Biofiltration planters offer high capacity for stormwater water quality treatment and peak flow rate reduction, and can be sited flexibly in contexts where infiltration is not possible or desired.

CONSIDERATIONS

The underdrain pipe may be directly connected back to the storm sewer system or may have an orifice control structure (e.g., located within a catch basin or manhole) at the end of the pipe to regulate the rate at which water flows out of the system. If a structure is installed at the end of the underdrain pipe, consider also using it as the overflow structure if sited within the planter or locate it outside of the planter for ease of maintenance access.

With the added layer of an underdrain pipe, the overall excavation for installation of the bioretention planter will increase, which may require more adjacent areas (e.g., sidewalk or roadway edge) to be restored in order to install the deeper planter walls.

Structured footings may be needed given sidewalk or street conditions to prevent lateral movement of the walls of the biofiltration planter.

CRITICAL

The soil and drain rock layers must be encased either by constructed walls (often concrete) or liners to prevent infiltration into surrounding soil.

1 Install a perforated pipe at the base of the facility to collect the treated runoff.

2 Use a raised drain or curb cut (as illustrated) to drain overflow back to the graywater system that exceeds the design rain event.

The planter must be designed to drain within 24 to 72 hours.

Integrated seating acts as a barrier around deep planters, **SEATTLE, WA**

RECOMMENDED

3 Maximize the surface footprint of the biofiltration cell, especially if multiple cells drain into a continuous rock layer below. Multiple surface expressions can be connected to maximize the bioretention footprint of a facility. Undersized surface cells may be at risk for increased erosion or large amounts of debris.

4 Provide a pedestrian cut-through (at least 5 feet wide) approximately every 20 to 40 feet to facilitate access to the curb. Cut-throughs may separate cells or be provided as ramps.

Use native plantings that are suitable for each site or microclimate, are able to handle seasonal flooding or drought, and require minimal irrigation or maintenance.

OPTIONAL

Discourage pedestrian trampling and reduce soil compaction by using low barriers or hardy vegetative ground covers.

Seating can be incorporated into the planter sides to improve the public realm for people. Review overall vertical drop from bench to bottom of facility. If drop exceeds 30 inches, review if bracing or fencing at back of bench may be needed.

Bioretention Swale

Bioretention swales are shallow, vegetated, landscaped depressions with sloped sides. They are designed to capture, treat and infiltrate stormwater runoff as it moves downstream. Swales are less expensive to build but use more space for infiltration and conveyance than planters, and can handle low to moderate flows of runoff.

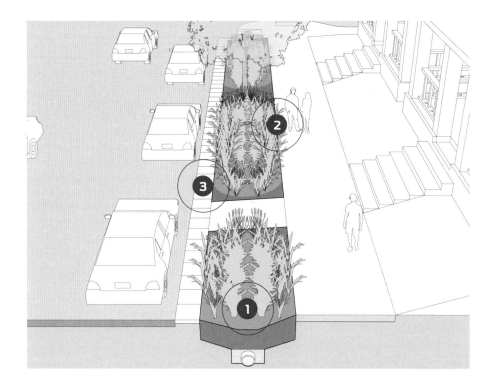

APPLICATION

Swales are most applicable in lower density or lower traffic contexts, as they have relatively large footprints and little or no vertical separation from the sidewalk and street. Swales are commonly implemented on neighborhood or residential streets, along shared-use paths, medians, roundabouts, or other unused right-of-way areas and in areas where more space is available for siting the facility within the planting strip or a curb bulb along the street.

Graded side slopes allow more flexibility in design and planting compared to bioretentions cells with fixed vertical walls.

On streets with excess asphalt—such as slip lanes or overly wide curb radii—swales can be applied as part of a strategy to reclaim space in the right-of-way for safety, livability, and stormwater management.

BENEFITS

Swales can support a wide range of plantings to increase beneficial habitat and greenscape. Swales also provide flexibility for planting a variety of street trees on the bottom, on side slopes, or at raised berms between cells.

The side slopes of a bioretention cell provide a transition from adjacent roadway and sidewalk grade to the cell bottom.

Graded side slopes are less complex to adjust or modify after installation than a fixed planter side, and may make access points to underground utilities easier to locate.

In most cases, bioretention swales are fairly shallow (generally less than 2 feet in depth) and do not pose major safety risks even in the event of incursion by people walking or by motor vehicles.

The level bottom area of a swale provides space for maintenance crews to stand while working.

CONSIDERATIONS

1 The bottom area of a bioretention cell along with the side slopes (up to the ponding depth) provides for the infiltration footprint area of the facility. More runoff can be detained as the flat bottom area is widened, or as the side slopes are made steeper. A minimum 12-inch bottom width is generally necessary, but may vary depending upon the bottom slope of the cell and the available space. In some cases, a minimum average bottom width of 18 inches might be a more appropriate criteria.

The degree of slope should be set considering the potential for erosion of the soil, operations and maintenance, whether the cell is sited in an area where people are more prone to driving vehicles off the road (such as near intersections or on higher-speed streets), whether or not a curb is present, and how much chance of walking into the swale is expected based on the adjacent pedestrian conditions and level of activity.

RECOMMENDED

Street runoff is directed into the swale through curb cuts, trench drains, pipes, or as sheet flow in curbless conditions.

The slopes of graded sides should be determined based on the potential for erosion, ability to establish plants, maintainability, and ability for pedestrians and people using mobility devices to recover should they enter bioretention facilities. Side slopes between 2.5–4H:1V (horizontal:vertical) are best suited to urban contexts, providing a balance between pedestrian comfort and landscape establishment. Vegetation and maintenance access are best supported by at least a 3H:1V or shallower slope.

Side slopes may be different on the sidewalk side and street side based on specific contextual requirements or concerns.

2 In areas where pedestrians may step off the sidewalk or curb into the bioretention swale, hardy vegetative ground covers can delineate the cell's edge. Barriers such as short fencing can also help protect the facility from trampling, similar to that found on urban tree pits. Regular pedestrian access paths between the sidewalk and street can also discourage entry into swales.

3 In areas with significant on-street parking turnover, consider paving a 12- to 24-inch concrete strip at the back of the curb. In this instance, tailor the side slope to 3H:1V for the reduced level area in the step-out zone for pedestrian comfort.

Level areas at the edge of the bioretention swale should be compacted to support people walking on the surface as well as vehicles and people using bikes, strollers, or wheelchairs who may encroach on the swale from the adjacent roadway or sidewalk. Consider using undisturbed, firm, native soil for the edge zone. If placing soil in this zone, consider compaction to 95% density in the edge zone.

OPTIONAL

In constrained contexts, 2H:1V side slopes or steeper may be acceptable. These facilities are typically shallower in overall depth (generally 12 inches) and are more susceptible to soil erosion if the plants do not establish well and stabilize the soil.

Local stone boulders and cobbles can provide aesthetic enhancements, increased vertical slope, and erosion control if they are well-embedded and substantially anchored.

The infiltration gallery can be recessed under a vaulted sidewalk to increase detention capacity.

Where swales are planted with grass or lawn, a 3H:1V or 4H:1V side slope will ease mowing and maintenance access.

The bottom width may be uniform, or may be tapered if the cell slopes down, creating a more uniform ponding area across the cell. (See page 103 for further detail.)

Additional slope protection or reinforcement may be needed for stabilization if planters are deeper than 12 inches.

Bioretention swale, **SEATTLE, WA**

Bioretention swale, **SEATTLE, WA**

A swale with stormwater trees doubles as space for snow storage, **MINNEAPOLIS, MN**

TYPICAL DIMENSIONS (NOT ILLUSTRATED TO SCALE)

PARKING-ADJACENT

18 - 36 IN STEP OUT · 12 IN MAX PREFERRED · 12 IN DESIRED MIN · 2.5 - 4H · 12 IN DESIRED · 1V

Swales adjacent to parking should provide a curbside level step-out at least 18 inches wide. Typical side slopes are between 2.5–4H:1V, though 2H:1V may be acceptable in shallow cells. The steeper the side slopes and deeper the facility, the more step-out space will be needed for comfort.

CURBLESS / FLUSH STREET

24 - 36 IN · 1V · 2.5 - 4H · 24 IN MAX · 6 - 9 IN TYP, 12 IN MAX · 12 IN DESIRED · .5 - 2%

Provide at least a 2-foot level area along all sides, with either low plantings or vertical warning elements to discourage accidental entry. Ensure the maximum ponding depth (page 102) is below the street/sidewalk grade to avoid flooding.

Hybrid Bioretention Planter

A hybrid bioretention cell combines elements of both swales and planters, featuring a walled side opposite a graded side slope to increase vegetated space and infiltrating area, while providing a softer streetscape treatment for people walking. Walls or graded sides can be configured adjacent to either a street or sidewalk, and can utilize a range of materials including concrete, rocks, or steel-faced curbs.

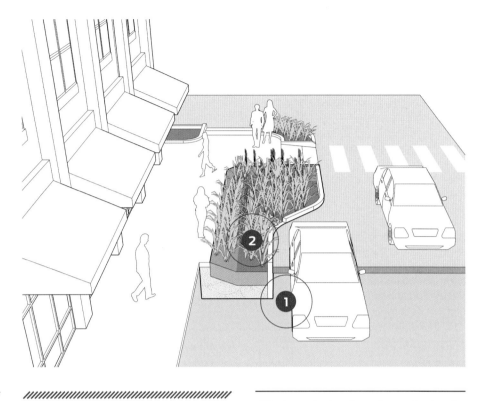

APPLICATION

Hybrid cells are used in low- to moderate-density, low-traffic contexts, and are typically placed in the planting strip on neighborhood or residential streets. Planting strips are often not wide enough to accommodate fully graded side slope bioretention swales, but are wide enough for hybrid cells.

BENEFITS

The hybrid biorentention planter can reconcile demands for space with a strategy for stormwater management, and can fit into a commonly available space on existing streets.

Graded side slopes are less complex to adjust or modify after installation, such as to retrofit for future utilities or service installation, compared with a fixed, difficult-to-modify vertical wall.

CONSIDERATIONS

The availability of space between the sidewalk and the curb may affect decisions about hybrid bioretention planter design. If space is constrained, locating the vertical wall on the sidewalk side may require less overall width for the facility footprint.

Design hybrid bioretention planters to accommodate street trees. The graded side slope is usually the best place for street trees, preventing the vertical wall from hindering tree growth. Additionally, avoid conflicts between location of subsurface utilities and planned root space for street trees. (See page 86 for more information on root space.)

Consider whether the desired tree canopy should shade the street or the sidewalk when deciding whether to site the planter with graded side slopes on the street side or sidewalk side. Plan for mature tree height and branching clearance—develop clear criteria for vertical clearance desired over bicycle facilities (typically 8 to 14 feet), pedestrian paths, or travel lanes.

1 In contexts where the vertical wall is located on the street side, the wall should be designed to support vehicular loads. This construction process may require new curbs and/or repaving for a portion of the street.

In contexts where the vertical wall is located on the sidewalk side, the wall should be designed to support pedestrian loads, and may require construction of structured footings, bracing, and potentially sidewalk replacement.

Where the vertical wall is located on the street side, the wall should be designed to support vehicular loads. This construction process may require repaving for a portion of the street.

RECOMMENDED

As with bioretention planters (page 80) and swales (page 82), a maximum 18-inch depth (from sidewalk grade to cell bottom) and 2.5H:1V side slope are typically preferred to balance detention capacity with adjacent mobility. Hybrid planters are typically designed to be shallower than walled planters due to the required width for the graded edge.

2 Three feet is a preferred minimum bottom width for stormwater performance, space consumption in the right-of-way, and maintenance access to the cell bottom. Width may vary along the graded edge, especially along a longitudinal slope to achieve greater infiltration area.

SE Division Street, **PORTLAND, OR**

TYPICAL DIMENSIONS (NOT ILLUSTRATED TO SCALE)

PARKING-ADJACENT

18 - 36 IN.
STEP OUT
12 - 18 IN. MAX
6 - 9 IN TYP, 12 IN MAX
36 IN. DESIRED MIN
12 IN. DESIRED
.5 - 2%
1V
2.5 - 4H

Hybrid planters adjacent to parking should provide a curbside level step-out area. Side slopes may be as steep as 2.5H:1V (2H:1V in constrained conditions), though 3–4H:1V may be preferred. The steeper the side slopes and deeper the facility, the more step-out space will be needed for comfort.

CURBSIDE TRAVEL LANE-ADJACENT

18 IN
DESIRED
1V
2.5 - 4H
12 - 18 IN MAX
4 IN
6 - 9 IN TYP

Planters adjacent to a curbside travel lane should provide an 18-inch level area from the curb face to the top of side slope to accommodate maintenance access.

CURBLESS / FLUSH STREET

24 - 36 IN.
1V
2.5 - 4H
12 - 18 IN DESIRED MAX

Provide at least a 2-foot level area along all sides, with either low plantings or vertical warning elements to discourage accidental entry.

Stormwater Tree

In addition to their immense social and aesthetic value, street trees provide quantifiable economic and ecological value to cities. Healthy street trees can contribute significantly to green stormwater management, with large capacity to transpire water, intercept rainfall, and treat water quality.

Boren Avenue N, **SEATTLE, WA**

APPLICATION

A tree well or pit is a box housing a single tree. Wells can have walled sides or structural soil systems to protect soil from compaction and retain stormwater.

Tree trenches are connected or linear tree boxes that usually have a subsurface system for distributing runoff among a series of trees. Tree trenches are often constructed in the sidewalk furnishing zone, though may also succeed in center medians or service street medians.

BENEFITS

Street trees can contribute significantly to green stormwater management by absorbing rainfall, transpiring water, and controlling runoff.

In cities, the high percentage of developed land and impervious land cover, including asphalt and concrete, contributes to an urban heat island effect: temperatures in a city can be significantly higher than in surrounding rural areas. Street trees can help mitigate the urban heat island effect through evapotranspiration and shading.

Urban air pollution, largely attributable to vehicle emissions and fossil fuel combustion, poses a public health threat, exacerbating respiratory and cardiovascular disease. Street trees play an important role in improving local air quality, removing air pollutants and filtering particulate matter.

Street trees calm motor vehicle traffic by visually narrowing the street and providing a well-defined roadside edge. The presence of trees can reduce speeding and crashes, improving safety for all street users.

Street trees play an enormous role in a street's livability, creating shade, dampening noise pollution, improving mental well-being, reducing stress, adding aesthetic value, and contributing to a sense of place.

Street trees have quantifiable economic benefits. Building energy consumption (and therefore costs) are lower on streets with shade. Street trees have been shown to increase property values and sales at local businesses.[3] Tree shade can slow pavement deterioration, decreasing maintenance needs and costs.

Urban forestry is an important strategy for climate change mitigation and adaptation. Street trees capture and store carbon dioxide and, by reducing temperatures and surface albedo, trees help with local adaptation to a warming climate.

CONSIDERATIONS

Planting street trees inside walled planters requires adequate cell width for a medium or large tree species to grow to maturity.

Smaller trees planted in depressed cells may have low branching structures that will overhang into the pedestrian path and create a barrier for people walking. Choose species based upon height clearance of branches and consider the expected need for pruning.

If the facility has an underdrain pipe, this may further inhibit the ability to place trees and provide clearance from the underdrain pipe and vertical wall. To make up for the lack of trees planted within the stormwater planter, consider widening the crossings between cells and review the overall street design.

CRITICAL

Design tree pits and trenches with adequate root space for the tree species underneath the sidewalk or street. As trees age, inadequate root space may degrade tree health and buckle pavement or sidewalks, creating hazardous, inaccessible, and expensive-to-repair conditions. Use root barriers to direct roots to appropriate growth area.

Plant diverse, climate-appropriate tree species that will be able to withstand the likely stormwater runoff and detained ponding depth, and that will support infiltration and transpiration. Consider the regional climate and expected future climate changes when selecting tree species. In winter climates, select salt-tolerant tree species and consider the need for snow storage.

For trees in the sidewalk furnishing zone performing bioretention, provide an appropriate inlet to capture runoff and distribute to tree boxes. This may be a curb cut or depression, or a catch basin that circulates water through connected tree boxes using capillary action.

When trees are adjacent to the bikeway, ensure branches do not impede bicyclists. Branches that overhang the bikeway or street should hang no lower than approximately 8–14 feet (depending on city requirements) to the bikeway or street surface.[4]

RECOMMENDED

Plant diverse, climate- and region-appropriate tree species. Varying trees can provide food and habitat to different species of birds and wildlife, and can increase resilience to different kinds of disease. Avoid over-specialization of a limited number of tree species, and consider using some evergreen and coniferous trees to maintain year-round benefits.

Select tree species that provide adequate shade and canopy, especially where people are expected or desired to sit or use the public realm.

Stormwater tree well, **LOUISVILLE, KY**

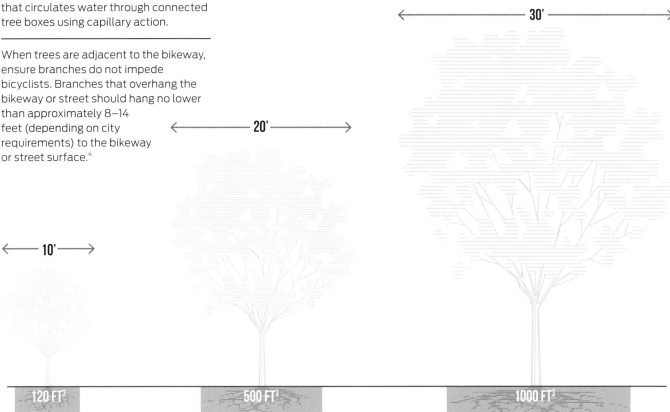

Larger trees require greater volumes of uncompacted soil space for roots to grow; though requirements vary by species, a tree with a canopy 30 feet wide needs roughly 1,000 cubic feet of root space to thrive.[5]

Permeable Pavement

The high amount of impervious surface cover in cities is a fundamental contributor to urban stormwater challenges. Decreasing the amount of impervious surface cover through the use of permeable pavement materials allows water to infiltrate through streets and sidewalks, reducing runoff.

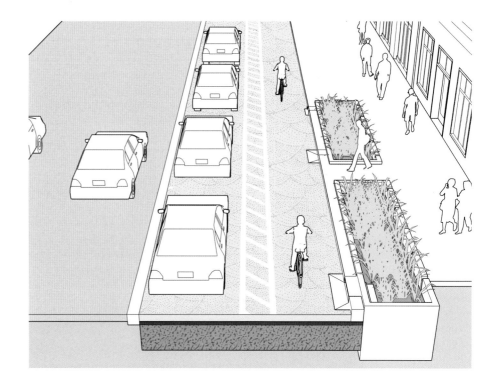

APPLICATION

Porous ashpalt, pervious concrete, permeable interlocking concrete pavers, and grid pavers provide infiltration of stormwater directly under the street surface, and they can be applied on any portion of the street provided appropriate surface and subsurface conditions.

Pervious pavements are most often applied on bikeways, parking lanes, and streets with lower vehicle traffic volumes and limited heavy vehicles. Sidewalks may also employ porous concrete to increase infiltrative area.[6]

BENEFITS

Permeable pavements add space for infiltration, especially in contexts such as alleyways or against the curbline, where nuisance flooding and other issues must be addressed without sacrificing space for mobility.

Many different types of pervious surface can be applied to different contexts, adding quality and performance to the urban environment.

CONSIDERATIONS

Pervious treatments must be carefully selected and installed for context and expected usage. Permeable pavements are often not appropriate for travel surfaces with high volumes, heavy vehicles, or where frequent starting and stopping place additional force on road surface.

Interlocking pavers may settle or buckle, reducing surface smoothness and creating accessibility issues.

Review adjacent surfaces and assess where soil is more likely to erode or be tracked onto the permeable pavement surface (such as from gravel driveways on private property). More frequent street vacuuming of those areas may be required to remove sediment.

Permeable street surface requires regular maintenance to remain effective. Develop detailed plans and responsibilities for appropriate cleaning and maintenance of permeable pavements installed in the right-of-way, which may include sweeping, washing, or vacuuming to remove grease, oil, and other sediment.

Permeable pavements are most effective on shallow slopes (5% or less) to allow water to infiltrate and for efficiency in spacing temporary subsurface storage.

Assess operations during snowy conditions. Avoid placement of snow piles onto permeable pavements since those are areas where higher concentrations of sediment will occur, thus increasing maintenance cleaning of the permeable pavement surface. Consider integrating alternative plowing and de-icing techniques to maintain the longevity of the infrastructure.

For interlocking pavers, aggregate used between paving blocks may need to be periodically replaced as it becomes loose or is pulled from between blocks.

Porous concrete parking lane, **VENTURA, CA**

RECOMMENDED

For pervious bicycling surfaces, apply porous asphalt or concrete rather than interlocking pavers, which may settle over time and become much less comfortable for bicycling.

If the bike lane is adjacent to a curb, consider not using a gutter and pave the permeable pavement up to the curb in order to avoid water flowing down the gutter and bypassing the permeable pavement. The curb can also be extended in depth to act as a barrier for deterring lateral movement of water once it has passed through the surface layer.

People riding bicycles usually prefer smaller voids in the porous asphalt and pervious concrete as riding surface, which provide smoother riding surface. Select a mix that has a lower water content, cementious ratio, and smaller aggregate.

Review the amount and characteristics of run-on flowing onto the permeable pavement system for the whole roadway (not just the parking or bike lane) and adjacent areas. Run-on, especially from areas with erosive soils, will increase the maintenance frequency for cleaning the pavement if the source of the sedimentation is not addressed.[7]

While bike lanes are not pollution generating surfaces, if the bike lane receives run-on from adjacent pollution generating surfaces (e.g., parking or travel lane), review local codes for regulatory requirements related to water quality treatment. In some cases, a water quality treatment soil media layer may be required within the pavement section prior to stormwater infiltrating into the subsurface native soils.

For more detailed technical guidance on the design and implementation of pervious paving treatments, refer to ASCE's *Permeable Pavements Recommended Design Guidelines*.

Porous asphalt increases infiltration on a small street, Percy Street, **PHILADELPHIA, PA**

Porous asphalt in a raised bikeway increases contrast from the pedestrian zone, and mitigates icing during winter months, Western Avenue, **CAMBRIDGE, MA**

Permeable interlocking pavement, Cultural Trail, **INDIANAPOLIS, IN**

Case Study: Southeast Atlanta Permeable Paver Roadways Project

Location: Atlanta, GA

Street Context: Residential Streets

Project Length: 4 miles

Right-of-Way Width: varies

Participating Agencies: City of Atlanta Departments of Watershed Management and Public Works

Timeline: Planning started 2012, Construction 2015–2016

Cost: $15.8 million, including $1.1 million utility allowance and 3 years of maintenance costs

GOALS

Flood mitigation: Manage the rainfall of a 25-year, 4-hour rain event akin to the storms in 2012 that caused neighborhood flooding.

Stormwater management: Reduce the volume of stormwater entering the combined sewer system, which serves about 10% of the city.

Infiltration under permeable pavement in the roadbed is complemented by curbside bioretention planters connected in the subsurface, Connally Avenue SE, **ATLANTA, GA**

OVERVIEW

After back-to-back storms caused severe flooding in the Atlanta neighborhoods of Peoplestown, Mechanicsville, and Summerhill in 2012, the City dedicated funds to improve the area's stormwater management. The neighborhoods, which are more than 60% covered by impervious surface, fall at a choke point within the combined sewer area, so larger storms caused neighborhood flooding and CSO events. City staff studied the entire sewer shed to map drainage patterns and define the project area and goals.

Four miles of streets within the project neighborhoods were milled, excavated, and resurfaced with permeable pavers atop an infiltration reservoir. This design allows stormwater runoff to infiltrate through the street surface, mitigating flooding and reducing overflow events.

DESIGN DETAILS

Typically, new developments are required to use green infrastructure to reduce the volume of runoff generated from the first inch of rainfall. This project had more robust goals: manage the rainfall from a 25-year, 4-hour storm event, or about 3.68 inches of rain.

The project scope included the installation of approximately four miles of permeable pavers within the flooding area and at the head waters of the drainage system. Though the original design called for six miles of permeable pavers, utility conflicts limited opportunities on two miles of streets. Permeable pavers have not typically been used in the US for an entire roadway, but the goal of this project was to manage as much volume as possible.

Designers maintained a gentle crown on the roadway so that water would still funnel toward the curb in the case that the entire permeable surface was clogged. Below the permeable pavers, 3–4 feet of aggregate provide the required storage volume to manage the design storm plus structural support for streets with vehicle traffic and bus routes.

Crumley Street SE, **ATLANTA, GA**

In addition to the permeable pavers, the City of Atlanta integrated about 30 stormwater planters to intercept runoff from the sides of the roadway. The soil within the planters rests atop a thin layer of gravel that is directly connected to the gravel reservoir underneath the permeable pavers on the roadbed; this design allows wicking to keep the vegetation in the planters watered during smaller storms.

An impermeable liner was installed on the sides of the streets to prevent water seepage into private properties.

Some steeper streets within the project area—up to 13% grade—posed a challenge for permeable pavers, which are easier to install on flatter streets. The design used impermeable liner check dams every 15–20 feet to break up the flows and allow better absorption into the subsoils on steep grades.

KEYS TO SUCCESS

Use a design-build contract that includes maintenance. The City developed 60% of the design drawings before involving a design-build contractor. The contractor is also responsible for three years of maintenance, covering tasks like repairing cracked pavers and performing street sweeping. After three years, the City will assume maintenance responsibilities.

Construction crews place pavers over a 3- to 4-foot aggregate layer, **ATLANTA, GA**

LESSONS LEARNED

Communication is key. The project required closing private driveways for short periods during construction, so the Department of Watershed Management required the contractor to have a public information officer on-site at all times while work was being done.

Expect utility conflicts. Utilities, streetcar tracks, and gas lines presented unexpected complications to the project. Although the project scope needed to be modified, a design-build contract improved the City's ability to work through unexpected utility challenges. Assess the existing utilities, protect those that are in good condition, and budget time and money in the project plan for any necessary utility replacement or relocation.

Identify maintenance hot spots. The City pinpointed specific places within the four-mile project area that need extra attention. Regular street sweeping is performed with a regenerative air sweeper.

OUTCOMES

In addition to managing stormwater runoff, the permeable pavers, along with bumpouts and planters, provide a traffic calming element.

Flow monitoring equipment and infiltration wells were installed in certain locations to assess the performance of the system. Though it is too early to evaluate results, preliminary findings show improved combined sewer flows in the system resulting in less neighborhood flooding.

WASHINGTON, DC

Green Infrastructure Configurations

As new design tools and best practices have emerged to improve multi-modal safety and accessibility on urban streets, new spaces have been revealed where green infrastructure can achieve both environmental and mobility goals. Apply special configurations for green infrastructure to compound the benefits of traffic calming and streetscape improvements.

Stormwater Curb Extension

Curb extensions visually and physically narrow the roadway width, creating safer and shorter crossings for pedestrians and providing traffic calming on low-speed neighborhood streets and commercial corridors. The available space generated by curb extensions can be used for bioretention, plantings, street furniture, benches, and street trees.

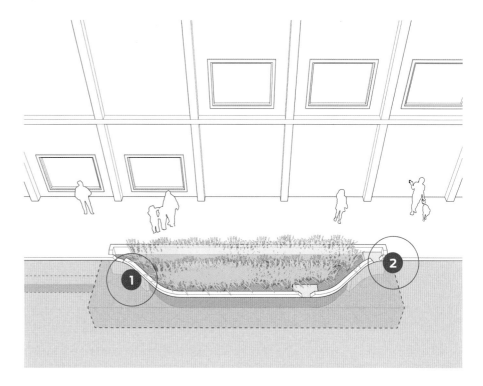

APPLICATION

Curb extensions—sometimes called "bulbs," "bump-outs," or "neckdowns"—can be sited at intersections to narrow the crossing distance for pedestrians and decrease curb radii, encouraging drivers to make slower turns.

Bioretention cells can be integrated into intersection curb extensions depending upon drainage patterns. Plantings can be used at corners to visually connect crossings, with or without bioretention.

Midblock curb extensions can be used to intercept and infiltrate gutter flow while providing traffic calming benefits; midblock bioretention bulbs can be combined with midblock pedestrian crossings.

Curb extension bioretention facilities are typically applied as walled cells, though hybrid cells (see page 84) with the graded side along the sidewalk edge can provide a gentle transition.

BENEFITS

It is relatively simple to direct gutter flow into curb extensions, especially when sited at the downstream end of the flow path.

Installation of stormwater curb extensions at intersections can create opportunities to improve accessibility and pedestrian safety. Shortening crossing distance, tightening corner radii, upgrading curb ramps, and relocating obstructions such as signal poles and utility boxes can improve pedestrian mobility and reallocate impermeable space for green infrastructure.

Midblock and intersection curb extensions that pinch traffic can reduce motor vehicle speeds, increase pedestrian prominence at crossings, and improve safety and comfort.

CONSIDERATIONS

Stormwater curb extensions should be sited to intercept runoff before it reaches gray drainage infrastructure. Location of existing catch basins may dictate placement or ability to implement intersection curb bulbs, especially if the catch basin is located at a corner apex or in close proximity to a pedestrian curb ramp. If catch basins cannot be moved and preclude placement of a stormwater curb extension at the intersection, midblock curb extensions can be sited immediately upstream of storm structures to intercept as much runoff as is possible.

On streets with minimal slope, the low point along the curb bulb may not be in line with the gutter. Review the street's longitudinal slope and cross-slope to site inflow points.

If runoff exceeds bioretention cell capacity during large storm events, overflow (or gutter spread) from the curb extension may extend into the roadway, reducing visibility and creating hazards from street users.

CRITICAL

Use low plantings in bioretention facilities and landscaped curb extensions near intersections to maintain sight clearance; plants should grow no higher than 24 inches above the sidewalk grade.

RECOMMENDED

The width of a curb extension varies with street type. Curb extensions are typically recessed 1–2 feet from the outside edge of the right-most travel lane, though width may be tailored to accommodate emergency or large vehicle access, or other existing conditions.

1 The curb return from bump-out edge to original curbline should be designed to enable street sweeping along the curb edge, typically angled between 30 and 60 degrees relative to the curbline. Steeper return angles will usually require hand-sweeping.

2 Design inlets and outlets to resist incursions by vehicles and bicycles, as motor vehicle wheels may be prone to enter, especially during parking maneuvers. Metal lids are an effective design strategy to block vehicle entry.

Inlets should include a presettlement treatment, especially along the curbline where runoff flows straight from the gutter (see page 108).

No lateral offset distance is required along the edge between the curb and bioretention area, as parking is not permitted adjacent to the curb extension. An 18- to 24-inch level area may be included along the curb to ease maintenance access from the street.

OPTIONAL

Utility corridors often determine where cells can be installed. Utilities may be sleeved, bioretention cells can be set back, or curtain liners may be integrated into design to prevent loading of infiltrated runoff onto subsurface infrastructure.

The bioretention cell helps to create a safe pedestrian crossing, **HOBOKEN, NJ**

The intersection corner area may be used for plantings or even bioretention area, though designers must carefully balance project goals and uses. Corner plantings increase the lateral offset of crosswalks, directing pedestrian paths around planted area and potentially decreasing visibility for turning vehicles.

Locating curb extensions in areas where on-street parking is already prohibited, such as near fire hydrants or driveway setbacks, can enable GSI implementation without affecting parking or curbside access. Ensure emergency responders have necessary access to utilities at all times.

Walled bioretention cells may incorporate seating and furniture around edges.

Midblock stormwater curb extension, **PHILADELPHIA, PA**

Intersection stormwater curb extension, **PALO ALTO, CA**

Stormwater Transit Stop

Bioretention cells incorporated into bulb-style transit stops improve the passenger waiting experience, but require special consideration to maintain accessible loading and clear sightlines for transit operators.

Incorporating stormwater facilities at transit stops introduces new opportunities for mutual benefits and interagency collaboration, unlocking new project funding sources and leveraging complementary resources.

Refer to the *Transit Street Design Guide* for further guidance on transit stop design.

APPLICATION

Bioretention cells can be integrated into transit bulbs at either end of the transit boarding platform, though are typically implemented at the end farthest from the intersection so as not to conflict with the path to the crosswalk,

Alternately, bioretention cells can be installed between the transit boarding platform and the pedestrian through path.

BENEFITS

Curb extension planters installed on transit bulbs can increase comfort for people waiting. Transit riders perceive much shorter wait times when greenery is incorporated into transit stops.[1]

Transit bulbs enable faster and more efficient transit stops and loading, allowing the bus to stop in-lane rather than pulling from and re-merging with traffic. Transit bulbs are also more space-efficient than pull-out stops, requiring less linear curb length by eliminating bus transition distance and freeing space for other curbside uses (including, but not limited to, green infrastructure).

Extending the boarding area out from the sidewalk reduces congestion and conflicts between moving and waiting pedestrians. Green infrastructure can further delineate sidewalk space between the pedestrian zone and transit station space.

CONSIDERATIONS

Tree canopy increases comfort for riders, but must not impede transit vehicle or pedestrian access to the platform.

Transit platforms may have higher curbs and platforms than typical sidewalks to accommodate level boarding (12–14 inches) or near-level boarding (8–11 inches), especially at streetcar and light rail stops. Raised boarding platforms require accessible ramps and access paths from the sidewalk, and will affect how runoff is directed into integrated green stormwater facilities.

Any green expression at a transit stop must be carefully coordinated with the transit operator to ensure transit vehicles and users can access the boarding platform easily and comfortably.

CRITICAL

1 At minimum, a 4-foot clear path must be available from the pedestrian through zone to any transit door, as well as into transit shelters and to access any transit amenities (e.g., ticket vending machines, maps, and wayfinding).

2 A solid, stable boarding pad that is 5 feet wide by 8 feet deep must be accessible to at least the front door to accommodate deployment of bridgeplates or ramps for passengers using wheelchairs.[2] Ten-foot-wide boarding pads are often preferred by the transit operator to allow flexibility accessing the stop. Any cells or plantings must not impede accessible paths.

Tree branches and plantings must not block transit vehicles or sightlines approaching the platform. For the approach side of the stop, select low-growing vegetation or trees where branches will not conflict with approaching vehicles.

RECOMMENDED

Coordinate with transit operators during the design phase to ensure that design and configuration of stormwater facilities do not conflict with the current transit fleet, or potential future operational changes. If a corridor may be served by enhanced service in the future, design transit stops flexibly to serve multiple vehicle types, or even multiple transit vehicles simultaneously (see right middle photo).

Provide rails between bioretention cells and transit platforms; in addition to protecting plantings, rails provide space for waiting passengers to lean or rest.

Alternatively, low curbs or fencing may provide adequate delineation for pedestrians around shallow cells; where cells are deeper than 6–8 inches, consider more robust edge designs that prevent accidental entry.

Inlets should have lids or covers where the curb is not parallel to the roadway to prevent vehicle incursion into the planter cell. (See page 106 for additional inlet design guidance.)

A stormwater planter at the end of a boarding island, SW Moody Avenue, **PORTLAND, OR**

Stormwater planters configured behind the boarding area are connected to a catch basin on the curb, collecting street and sidewalk runoff, Bedford Avenue, **NEW YORK, NY**

Bioretention on a bus bulb, SE Division Street, **PORTLAND, OR**

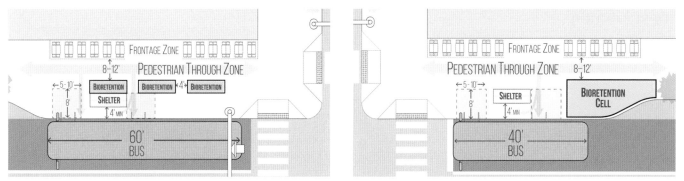

Maintain accessible clear paths at least 4 feet wide around planters, shelters, and seating. Site bioretention facilities clear of pedestrian through zone, and ensure trees and vegetation remain clear of transit vehicles.

Floating Island Planter

Vegetated space on pedestrian refuge islands, bikeway buffers, transit boarding islands, or other constructed elements that are offset from the curb can improve the streetscape and provide space for street trees. Bioretention planters may sometimes be incorporated into islands, provided suitable topography and thoughtful design.

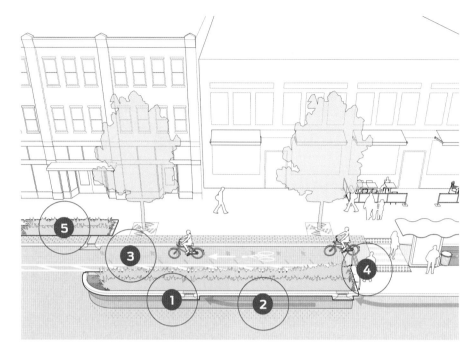

APPLICATION

Due to space considerations, walled bioretention planters are typically most applicable on "floating" street design elements.

BENEFITS

Providing green, vegetated space on pedestrian refuges, transit islands, and bikeway buffers can enhance the quality and aesthetic of transportation infrastructure that confers strong mobility benefits. Elements that calm traffic, shorten crossing distances, and protect people bicycling can open new opportunities for trees and vegetation that enhance the streetscape.

CONSIDERATIONS

Calm motor vehicle speeds and maximize visibility of floating elements to limit the risk that someone will drive a motor vehicle into the bioretention facility.

CRITICAL

1 Runoff is directed into the island planter either as sheet flow or using trench drains, gutter pans, or linear curb to efficiently collect stormwater.

2 Overflow runoff must be managed along the curb as well to prevent pooling in the curbside lane.

RECOMMENDED

While raised curbs are recommended to protect against vehicle incursion into planters, curbs with inlet cuts or depressions may gather less runoff than curbless stormwater planters. If a curb is not installed, consider using vertical reflectors or bollards to discourage vehicle entry into planters.

3 Maintain at least 5–8 feet between a floating planter and the curb to allow sweeping and maintenance between elements. Special sweeping equipment may be required for lanes under 8 feet.

Use low plantings to maintain sightlines and visibility, especially where pedestrians enter or exit the island.

4 On transit boarding islands, railings provide space for transit riders to lean while waiting and also prevent people from entering bioretention facilities.

5 Where a constructed buffer exists between travel lanes and a bikeway or curbside lane, install curbside bioretention to reduce pooling against the curb. Floating and curbside bioretention elements may be designed to work together—any excess runoff from the island planter is directed to a downstream curbside facility to reduce loading on both facilities.

Trees may be incorporated into transit islands and pedestrian refuge islands if width is available for the root space, provided branches will not grow to impede travel lanes or reduce visibility.

Rosemead Boulevard, **TEMPLE CITY, CA**

Stormwater Median

Wide medians used to separate traffic directions may be utilized for large amounts of water conveyance and infiltration. On very wide streets and parkways, green stormwater infrastructure can be coupled with greenways for bicycling and walking, providing attractive public space adjacent to stormwater management.

21st Street, **PASO ROBLES, CA**

APPLICATION

Bioretention swales, planters, and biofiltration planters can be integrated into medians where width allows.

Uninterrupted linear street space may manage large volumes of runoff from adjacent pavement or areas upstream collected from multiple blocks.

BENEFITS

Median bioretention facilities utilize potentially unused space in the right-of-way, and can repurpose that space into multi-functional street space by providing greenscape and stormwater infrastructure.

Median cells may be very high capacity stormwater facilities (depending upon right-of-way space), as access is more easily limited.

CONSIDERATIONS

Medians tend to be at the high point of a road's cross slope; thus, street runoff is typically not sheet flowing to a median. As a result, bioretention facilities sited in medians may require reversing a street's cross slope in order to intercept the adjacent street's runoff.

If the street's cross-section cannot be modified, review opportunities to intercept stormwater collected in an upstream conveyance system (from other streets) and having it daylight into a series of bioretention facilities in the median.

Sewer mains and other underground utilities tend to be located in the middle of a roadway and may need to be relocated.

Consider street sweeping, plowing, and snow storage in colder climates. Snow that is not removed may be stored in the median, which can impact plant health and facility durability. De-icing chemicals, salt, and sand may also impact tree and vegetation health, especially if applied heavily.

CRITICAL

Use low plantings in bioretention facilities and landscaped medians near intersections and at pedestrian crossings through the median to maintain sight clearance; plants should grow no higher than 24 inches above the pavement surface where pedestrians will gather or cross the intersection.

RECOMMENDED

Design the roadway to slope toward the median bioretention facility. This may include either a street with an inverted crown (all runoff flows to the center) or a "thrown" street, where the roadbed slopes in one direction, and the median intercepts half of the runoff.

Review maintenance operations during the design phase. Determine if lanes will need to be closed for maintenance crews to safely work within the median and clean inflow points into the facility, or if refuge areas can be designed into the median for maintenance access.

Provide an 18- to 24-inch-wide level area to ease maintenance access along either edge.

Trees can be planted in median stormwater facilities to manage runoff. Ensure that tree trunks and root structures are not submerged in water longer than the species is able to tolerate. Trees may be planted on berms to limit ponding around the base. Select species and place trees such that they do not impact visibility or street lighting.

OPTIONAL

Pedestrian walkways built around bioretention may provide an inviting or distinctive public space anchored by greenscape, especially on wide boulevards and parkways where large medians define the street context and where green infrastructure provides a visual cue to drivers to reduce speeds.

Argyle Street, **CHICAGO, IL**

Bioretention Design Considerations

Cells designed for bioretention can manage the stormwater runoff from large areas of impervious surface. Site and design bioretention cells with sensitivity and consideration for local context and the performance needs of each project within the stormwater network. Consider maintenance needs from the beginning of the design process, and include transportation and mobility goals alongside ecological goals.

Bioretention Cell Sizing

The length-width-height dimensions of a bioretention cell determine the cell's capacity for temporarily storing stormwater while it gradually filters through the cell's soil and either infiltrates into the native soil below or is collected in an underdrain pipe and discharged into the downstream system. How large a cell needs to be depends on the stormwater management goals for the facility and the space available in the right-of-way.

A cell's footprint area depends on the cross-section type for the bioretention facility and the length, width, longitudinal bottom slope, and planted area of the cell. For facilities with vertical sides, the footprint area is calculated as the bottom area of the cell plus the width of the vertical sides. For facilities with graded side slopes, the facility footprint area includes the bottom of the cell along with the footprint of the area on the sides when the cell is at its maximum ponding depth plus freeboard.

SIZING AND SITING CELLS

Bioretention cells are sized and modeled with consideration for factors including the amount of runoff the system is expected to filter downward through the facility's soil layers (and either infiltrate or collect in an underdrain system) and the amount of space in the street cross-section available for siting green infrastructure. Cell size is also informed by the street context, including factors like the extent of impervious surface area, adjacent land uses and structures, pedestrian volumes, and surrounding topography.

The size of stormwater facilities should account for regional conditions such as climate and weather patterns and local subsurface soil conditions and properties. Cells are typically designed to meet a drainage target of typical storms during a year; during less frequent but more intense rain events, such as a 25-year storm, cell design may allow intense flows to partially bypass the bioretention facility. If a region receives consistent but slow rainfall, and the site has fast-infiltrating native soils, smaller bioretention cells can provide water quality treatment and capture most runoff. Where large storm events are likely, even if infrequent, large bioretention cells or conveyance to large stormwater detention facilities may be desired to reduce peak flow into the graywater system.

Watershed and existing drainage basin systems affect cell design and size. In regions with combined sewer systems, the primary purpose of cells is generally to capture and infiltrate as much volume as possible to reduce overflows; a large bioretention cell capacity is needed to achieve this goal, usually resulting in a relatively large footprint. In areas with storm systems that have capacity to convey stormwater flows to an open water body such as river or lake, the primary purpose of the cells may be to capture and treat stormwater runoff to protect the region's water quality, but it is not always necessary to infiltrate all the stormwater for smaller, more-frequent storms.

Bioretention planter cells have vertical walls, providing more capacity for temporary water storage than cells with graded side slopes in the same footprint. However, there are other tradeoffs relevant to the decision to use vertical sides or graded side slopes at a given site.

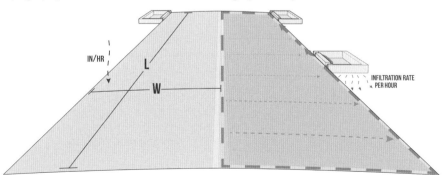

Bioretention cells are sized to detain and infiltrate an expected amount of rainfall on a specified street area and over a specified amount of time, typically bisected by the roadway crown.

Where space in the street cross-section is scarce and curbside access is a relatively low priority, long and narrow cells can achieve a high stormwater management capacity. Conversely, where curbside access is a high priority, short and frequent cells can provide a high stormwater management capacity while preserving pathways for people on foot.

To simplify bioretention cell sizing and modeling, cities and regions can develop local sizing charts based on simplified stormwater models. These simplified charts specify the percentage bottom area needed to drain an effective impervious surface area, given a design infiltration rate based on native soils—a useful tool within a reference region.

KEY CRITERIA

For both vertical walled and graded bioretention cells, a single cell's or cell system's ability to infiltrate runoff effectively relies upon three criteria:

> Cell wetted area footprint;

> Ponding depth; and

> Infiltration rate of underlying and engineered soils.

Bioretention Cell Wetted Area

provides the infiltration footprint area of the facility; the wetted area is the surface area at maximum ponding depth of a facility. A larger wetted area maximizes the infiltration area used in sizing the facility. As the bottom area increases, the wetted area increases to maximize the cell's storage capacity.

Engineered soils can increase the bioretention capacity of facilities, providing additional storage.

1 For **bottom width**, 4 feet is a typical preferred minimum for bioretention planters with walled sides; narrower planters may be possible, but these increase the challenge of maintaining healthy plants and are usually less cost-effective to implement, given construction and maintenance costs and performance. For bioretention swales, the minimum recommended bottom width is 1 foot.

2 **Cell length**, or the length of a bioretention swale or planter along the curb, can range from 10 feet to the length of an entire block with intermittent berms. Besides stormwater storage and infiltration capacity, cell length is also affected by longitudinal slope and design infiltration rate of bioretention soil media and native soils.

Aside from managing stormwater runoff efficiently, the length of an individual cell is driven by all-mode curbside access needs. Where parking is permitted, provide for sidewalk access from the curb about every 40 feet, or approximately the length of two parking stalls. Cell crossings can be provided at sidewalk level or may slope down to street level, but they must be above the ponding depth of the cell.

The full bottom length and width of an individual bioretention cell should be used for temporary stormwater storage; where possible, cell design should maximize temporary storage capacity.

A level bottom area accommodates relatively simple maintenance access for routine cleaning and plant care.

On streets with longitudinal slope, install elevated berms or check dams to allow runoff to pond and infiltrate downward through the entire cell bottom area, rather than only flow downstream to the end of the cell.

SE Tacoma Ave, **PORTLAND, OR**

3 **Ponding depth** provides for temporary storage of the stormwater before it filters downward through the bioretention facility. The temporary ponding depth for bioretention facilities ranges from 2 inches (for mitigating sidewalk runoff alone, or in fast-draining soils) to up to 12 inches (for mitigating roadway runoff, or in slower-draining soils). In areas with moderate to high pedestrian activity (e.g., commercial streets and business districts, and near busy community facilities), limit the ponding depth to 6 inches or install short fencing around the bioretention facility for depths greater than 6 inches.

Vector control issues make **drawdown time** and maintenance critical from a health and safety perspective. Typically, cells should drain out (have no surface ponding) within 72 hours after the rain event has ended to prevent insect-borne diseases (e.g., mosquitoes breeding cycle).[1] However, a 12- or 24-hour drawdown period is often preferable, especially in contexts with regular precipitation (to accomodate the next storm), high pedestrian volumes, or other influencing factors. If fast-infiltrating native soils are present, faster drawdown is also realistic.

4 **Freeboard depth**, measured from the maximum ponding depth to the top of the facility's overflow elevation, provides a buffer during larger storm events when water overflows the facility and flows into either a storm collection structure or back into the street gutter system through an inlet or curb cut. Freeboard depth is affected primarily by the street grade, and should range from 2–6 inches or greater, depending upon site context, potential for overflow occurrence and its impacts, capacity of the downstream conveyance, the volume of water that the facility is managing (runoff from one block versus multiple blocks), and other engineering or design judgements. Set a deeper freeboard (usually with higher walls) at locations with more-frequent overflows or greater potential impacts.

Inflow, Outflow & Overflow

There are three basic patterns for how water flows into and out of a bioretention facility, if it does not infiltrate through the bioretention facility into the native soil or into a subsurface underdrain pipe. These patterns are:

> **On-line/through-flow**

> **Combined inflow/overflow**

> **Raised overflow drain**

The design of the inflow, outflow, and overflow path depend on the amount of stormwater runoff that the facility is intended to infiltrate or convey, operation and maintenance requirements, site context, pedestrian and bicyclist safety considerations, and the type of bioretention cross-section.

Depressed curb inlet and outflow, **NEW YORK, NY**

DISCUSSION

On most streets with a curb, the curb should be continued along the bioretention facility edge. Accommodate water flow into the cell in one of four ways:

> a depression in the gutter and a break in the curb that directs the flow through a graded channel to the facility's bottom;

> an inlet;

> a catch basin with a pipe or rectangular channel daylighting into the cell;

> a depression in the gutter that connects to the concrete channel with a trench grate that then outfalls into the cell.

A cell along a curbed street may need one or more inflow points, depending upon the length of the cell, in order to maximize the amount of flow entering the cell and use the full capacity of the facility. Inflows should allow the cell to fill up to its full design ponding depth.

The outflow point serves to direct water back to the street's gutter or to another cell along the corridor. The inflow point may also act as the outflow point for the facility, depending upon the elevations of the street, inlet, and cell. Alternatively, a storm structure, usually a raised drain, may be placed within the cell to direct the overflow downstream when the cell reaches capacity, or a piped connection to the gray stormwater system for overflow volumes.

The overflow elevation is the maximum ponding elevation (see page 103); runoff must be directed through an outlet, overflow drain, or bypass and not enter the bioretention area to prevent street or sidewalk flooding.

CRITICAL

The opening and flow path into and out of the cell must be clear of obstructions. Drop the inlet grade 2 inches below the street grade, and allow another 1- to 3-inch drop behind the inlet to allow debris and sediment to settle without blocking the flow of runoff.

Locate other street infrastructure, such as posts for street signs, water meter boxes, valve boxes, and irrigation heads outside of the flow path if they are near inflow and outflow points.

Locate trees and shrubs at least 5 feet clear from the edge of the inlet to allow for tree/shrub growth without blocking flow path and ease of maintenance.

Review existing roadway and sidewalk surface grades and regrade or repair any uneven surfaces to ensure that overflow stormwater will flow from the bioretention cell into the street/gutter (or into an overflow drainage structure, if applicable), and not onto an adjacent property or the sidewalk.

Evaluate street grades to confirm that irregularities in the pavement surface do not prevent stormwater runoff or gutter flow from entering or exiting the bioretention facility. Sheet flow off the pavement can easily be unintentionally diverted via a crack or joint in pavement. In retrofit projects, it may be necessary to replace some street pavement to improve water flows from the crown to the gutter, and into the bioretention cell via the curb cut.

ON-LINE/THROUGH-FLOW

Stormwater flows can be designed to follow a path through multiple bioretention cells along a corridor, especially on streets with a gradual but consistent slope. Stormwater flows in the front (upstream end) of the cell, and excess volume above the designed ponding depth either flows out the back (downstream end) or bypasses the facility.

Through-flow design is most applicable where a series of cells manage stormwater along a corridor.

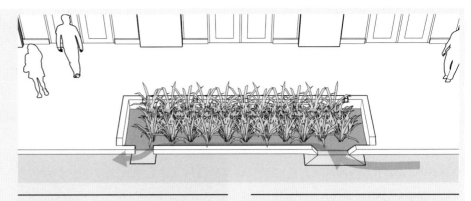

Enabling stormwater to flow through multiple cells allows the first cell in a series to have the presettling zone for targeted maintenance and then distributes water to other facilities.

On flat streets, to avoid water backing up in the street, the outlet elevation must be 2 inches below the inflow elevation. Otherwise, a raised overflow drain within the facility may be necessary.

COMBINED INFLOW/OVERFLOW

Bioretention cells can be designed so that overflow stormwater simply flows out of the same inlets through which it enters, especially along relatively flat grades.

Cells can be placed over an existing stormwater drain. Place one inlet on each side of the drainage grate to direct overflow into the gray system, with the catch basin at the low point in the street and directing all runoff up to the design volume into the bioretention cell first.

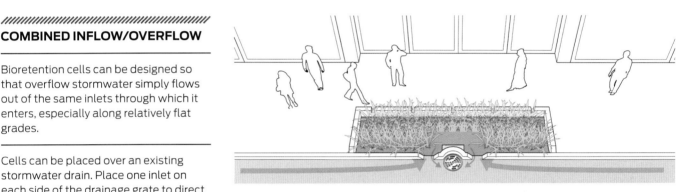

Combined inlets are applicable where each bioretention cell is infiltrating a discreet roadway segment, rather than acting as part of a chain or network diverting along a corridor.

RAISED OVERFLOW DRAIN

A raised overflow drain such as a "beehive" grate over a catch basin, area drain, or storm structure can be used as the outlet for overflow stormwater. These drains are designed to be level with the maximum ponding depth of the cell. The raised overflow drain should be placed with its rim elevation at the maximum ponding surface elevation.

Overflow is directed into the raised drain, and is piped down to either a stormwater chamber or gray stormwater system.

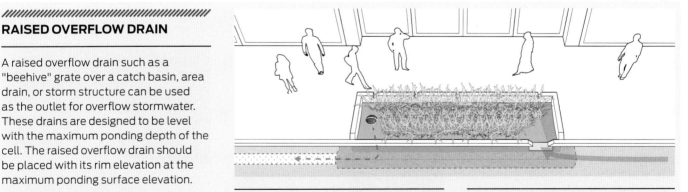

When using a raised overflow drain in a cell with vertical walls, provide adequate clearance between the outer wall of the structure and the raised drain to accommodate maintenance access.

The space that the overflow structure takes up does not count toward the infiltration area for the facility.

Inlet Design

Inlets convey stormwater runoff from the surrounding street catchment area into the bioretention facility. Details of inlet design—including width, grade, and location—guide the stormwater into and out of the stormwater cell.

DISCUSSION

The width and number of inlets determines the stormwater inflow and outflow capacity. Inlets should be wide enough to accommodate the expected stormwater volume, but their minimum size is usually related to the type of maintenance equipment that will be used to clean the curb cut (for example, the minimum width of a shovel). Inlets are typically 8–12 inches wide, but inlets up to 24 inches wide are not uncommon.

If there are multiple cells in a series or multiple inlets in a cell, some water can bypass the first inlet and then flow into the next inlet downstream.

Sequenced inlets, **SEATTLE, WA**

Building an inlet often involves minor modifications to the existing flow path in the gutter or pavement at the face of the curb. These changes can include a sloped surface through the curb cut to the cell to facilitate conveyance of stormwater into the cell, as well as a stabilized surface inside the cell at the inflow point (e.g., cobbles or a concrete pad) to minimize erosion as flow enters the cell.

Carefully evaluate the slope of the inlet to ensure flows can enter the bioretention cell. Small differences in slope may mean that low-flow runoff bypasses the inlet.

Depressed curb, **SEATTLE, WA**

Vertical notches can be installed along the sidewalk edge of the facility. A width of at least 4 inches will provide convenient maintenance access while minimizing the risk that people will get their feet caught in the openings.

Trenched inlet, **WASHINGTON, DC**

Design inlets to minimize the opportunity for entry by vehicles. Even narrow curb cuts can allow incursion, especially if perpendicular to the parking lane. Metal or concrete lids can reduce incursion risk.

Curb extension inlet, **SEATTLE, WA**

Design inlets to resist blockage. The accumulation of garbage, debris, or sediment at the inlet will prevent runoff from entering the bioretention cell, causing stormwater flows to bypass the cell and eliminating its value for flood prevention and runoff capture. Consider how ice and snow may block inlets and mitigate for this issue if necessary. Include a designated presettling zone—at the inlet for the first cell in a series, or at the primary inlet for a single cell on a block—for collecting debris and sediment, and to allow cleaning efforts to focus on a small space within each project.

Curb cut inlet, **VENTURA, CA**

Covered sidewalk inlets, **SEATTLE, WA**

Valley gutter, **CHICAGO, IL**

APRON/DEPRESSED GUTTER

PORTLAND, OR

Inlet aprons or depressions increase inflow effectiveness. Aprons typically drop 2 inches into the bioretention cell, with another 2-inch drop behind the curb to maintain inflow as debris collects.

Gutters with steep cross slopes can create hazards, especially to people bicycling. Curbside and protected bike lanes along concrete aprons should be at least 6 feet wide to give bicyclists adequate clear width from the curb and any pavement seams.

Where the curb alignment along the street is straight, the curb opening may optionally have a bar across the top of the inlet. For aprons into bioretention swales, the curb may angle into the cell to improve conveyance of gutter flow into the facility in the planting strip.

A depressed concrete apron can be cast in place or retrofitted in by grinding down the existing concrete pavement.

DEPRESSED DRAIN

SE Tacoma Street, PORTLAND, OR

In a depressed drain, runoff in the gutter drops into a grate-covered drain before flowing into the cell. Drain covers must be selected for compatibility with bicycling and walking; grid covers are preferred.

Depressed drains are a potential solution for bioretention cells on sloped streets where directing runoff into the cell is a challenge.

The curb is also fitted with an opening inlet to maintain runoff capacity into the cell in case debris collects on the grate.

INLET SUMP

Larch Street, CAMBRIDGE, MA

Where a large amount of debris is expected, an inlet sump can settle and separate sediment from runoff before entering the infiltration area.

Runoff drains into a catch basin, which collects debris in its sump. After presettlement, water drains via a pipe daylighting into the swale or through an opening in the catch basin's walls to the infiltration area.

Inlet sumps convey flow to the subsurface via a perforated underdrain pipe that distributes the flow to bioretention cells or connected tree trenches.

Avoid siting sump inlets where pedestrians will have to negotiate with them, such as at the curb return or immediately next to the curb ramp.

TRENCH DRAIN

Tilikum Crossing, PORTLAND, OR

A trench drain is a long, covered channel that collects stormwater and can direct it into the green or gray system. Trench drains are often used atop an existing gutter or stormwater path, and are an excellent solution for streets where walking across the entire surface is to be encouraged.

Trenches may either be shallow (where runoff volume is less of an issue) or deep and covered by a metal grate. Deeper trench drains may gather sediment and require frequent maintenance.

Trench drains may be configured either perpendicular or parallel to the flow direction of the roadway, collecting runoff and directing it to a single inlet in the bioretention cell.

Trench drains can be designed as detectable edges or part of a detectable edge, and may be used to help define flush or curbless streets, especially shared streets. Review current ADA code for grate requirements if trench drain is within sidewalk zone and/or ADA path of travel.

Presettling Zone

A presettling zone is an area made of cobbles or concrete that captures debris, pollution, and sediment at the upstream end of a bioretention facility. It prevents erosion in the facility, and provides a designated location for maintenance crews to target for cleaning debris and sediment.

Splash pad/flow dissipator collects debris from runoff entering the cell, **CAMBRIDGE, MA**

DISCUSSION

A presettling zone or structure serves to capture debris and sediment from the surrounding catchment area. A presettling zone is especially recommended in areas with concentrated point pollution discharges or significant sources of upstream sediment (e.g., exposed soil, automobile service and repair uses or industrial uses, or debris from tires). Designating a presettling zone or structure allows targeted maintenance in that area to remove sediment build-up.

The presettling zone should be designated within the first (upstream) portion of a bioretention cell. For multiple cells in a series on a block, the presettling zone will typically be in the first upstream cell at the inlet that receives larger flow volumes before it is distributed to the remaining cells down the block. Alternatively, use a presettling structure, such as a catch basin with a sump where the outlet pipe daylights into the bioretention cell.

The design of the presettling zone will vary depending on the tributary area characteristics such as amount of sediment load, street maintenance, traffic volumes and adjacent use in the right-of-way. Streets with higher traffic or with adjacent land use that is susceptible to erosion will generate more sediment than neighborhood streets with lower traffic volumes.

If the presettling zone includes a wider curb cut, review the potential for people to park a car with its tire in the curb cut. Some city agencies have used a continuous metal lid across the top of the curb in the curb cut to prevent vehicle incursions. With this approach, consider the maintenance equipment needs for clearing debris from the narrower opening.

If vehicular parking or loading is adjacent to the facility, provide a level step-out zone. If it is a critical inflow point, use no-parking designations to prevent cars from parking in a position that blocks the curb cut.

Consider the city's operations and maintenance equipment when designing the presettling zone, as well as the amount of impervious area footprint taken up by the presettling zone. For example, city agencies that use vactor trucks may prefer presettling zones with concrete pads rather than a rock or cobble pad. Other cities may prefer cobbles, which minimize the amount of impervious surface created in the cell compared to a concrete pad. Some city agencies may use a screen on the vactor hose to capture finer sediments without capturing cobbles. Some agencies prefer catch basins with sumps, given adequate space and depth to daylight pipes from catch basins into the bioretention cell.

Plant materials can be highly beneficial in the presettling zone to withstand inundation, soil accumulation, or erosive conditions. Plant materials provide a soft edge treatment, compared with cobbles or a concrete pad.

There is no straightforward method to size presettling zones. Streets with higher traffic or with adjacent land use that is susceptible to erosion will generate more sediment compared to neighborhood streets with lower traffic volumes. Design judgment with input from maintenance staff is important so that the presettling zone is within the context of the neighborhood character, street scale, and maintenance approach.

RECOMMENDED

Provide a scaled presettling zone at point discharges with concentrated flows. In most cases, a presettling zone should be located at the upstream end of the first cell in a series if the first cell is intended to receive at least half a street length of gutter flow.

Provide a presettling zone at inflow points on streets with higher traffic volumes, especially truck or bus traffic.

Where cobblestones are the desired material, use a mortar treatment to reduce the maintenance demands and replacement requirements of loose cobble.

Autumn maintenance leaf collection prior to anticipated storms is recommended if this is the primary inflow into the facility. See *Operations & Maintenance* on page 124 for more information.

2-inch drop settles light sediment, **PORTLAND, OR**

Concrete presettlement, **KANSAS CITY, KS**

Stone presettlement captures debris, **SEATTLE, WA**

Presettlement area with overflow structure, **AUSTIN, TX**

Stone forebay and step-out strip, **NEW YORK, NY**

Inlet sump and forebay, **AURORA, IL**

Check dam with rock on downside slope to prevent erosion, **PORTLAND, OR**

Soil, Media & Plantings

Selecting the appropriate soil media(s) and plant species is critical for successful bioretention facilities. Soil characteristics should be chosen to support drainage, pollutant removal rates, and plant health and viability. Plantings should include a diverse community of native species.

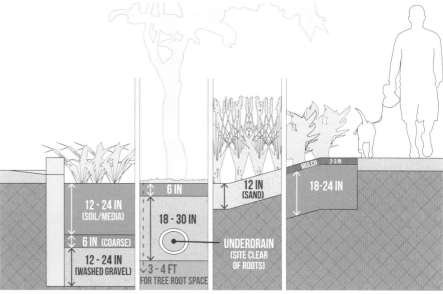

Media layers vary based upon design intent and context, with several possiblities shown.

SOIL MEDIA

The type(s) of soil media used in a bioretention facility varies based on region-specific conditions, water quality treatment requirements, flow attenuation, and regulations. The soil must:

> have proper permeability to drain;

> absorb the targeted pollutants and capture suspended solids;

> include the proper materials to avoid exporting pollutants such as nutrients or pathogens;

> sustain vegetation.

The characteristics of the bioretention soil media will affect the infiltration rate. Slow-draining media infiltrates less than 1 inch/hour, while faster-draining media can infiltrate 10 inches/hour or more.[2]

In general, the soil media(s) used in bioretention sections are considered non-structural, requiring engineers and designers to consider bracing for planters and/or setbacks from the roadway, sidewalks, and structures.

The layers of bioretention soil media may be limited to one type of mix, or multi-layered with two or three soil layers to target specific pollutants. Multiple or amended soil media layers may be used in areas with higher pollutant loads.

PLANTS & TREES

Plants and trees used in the bioretention cross-section provide a stormwater function and also improve the green, aesthetic experience in the right-of-way. Trees, plants, and their root structures help with absorption of stormwater, uptake of nutrients and pollutants, soil stabilization, and sustaining the bioretention soil media infiltration rate over the long term.

The type of plants and trees suitable for a bioretention facility varies across regions. When selecting plants, consider regional or local climate conditions and the species' ability to withstand varying wet and dry conditions. Plantings should include a diverse community of species appropriate for each specific site. Native plantings are preferred, but vegetation appropriate to each climate enhances the aesthetic and performance of the infrastructure. Consider salt tolerance in winter climates and flooding tolerance in regions with heavy storms.

The type of bioretention soil media will affect the type of plant species that can thrive. Media with higher concentrations of organics will allow for a wider selection of plants. Bioretention soil media that drains quickly (in cells with an underdrain or with native soils with high infiltration rates) may limit the types of native plants that can be used.

Plants and trees in green stormwater infrastructure can provide habitat for bees and other pollinators. Select a diversity of species to improve ecological health and minimize susceptibility to infestation or disease.

Consider tree placement carefully at the beginning of the project design phase. Trees should be sited adjacent to and/or within the bioretention facility so that adequate soil is provided for healthy long term tree growth. During the siting and selection of bioretention facilities, review the placement of trees along the street in coordination with other infrastructure and setback requirements.

Trees can be planted within the bottom of cells or on the side slopes of the bioretention facility. Consider the branching form as the tree grows to ensure that there is adequate height clearance on both the sidewalk and roadside.

When planting trees on the side slopes of bioretention swales, in order to have space for the tree pit and watering ring, the bottom width of the bioretention swale can narrow and the side slopes for tree planting can increase to 2H:1V to transition to the bioretention swale's sides.

Narrow bioretention planters or other facilities that are lined can limit the ability to plant medium to large trees, and smaller trees may not have a branching structure for sidewalk clearance. In cases where the bioretention facility type and/or location limits the ability to plant trees, consider widening the gaps between bioretention cells to make space for street trees, which help with stormwater absorption. Coordinate bioretention cell types and locations to optimize stormwater management and vegetation health.

Consider operations and maintenance needs and available resources when selecting soil media and plant species. Choose low maintenance plants that minimize the need for mowing, pruning, weeding, and irrigation. Avoid using fertilizers or pesticides.

CRITICAL

In bioretention swales with graded sides, vegetation should be planted in a contour pattern with regard to each plant's role in infiltration; plants at the lowest grade of the cell that will be submerged during storm events must be able to withstand flooding well, while plants along the graded side slopes must be effective at stabilizing soil and controlling erosion.

When selecting vegetation, review the plant species' mature height. For bioretention facilities near intersections, pedestrian crossings, and driveways, use a maximum mature plant height of 24 inches for sight clearance and visibility. If plants are planted in the bottom of the cell, the mature plant height can be greater so long as the height above street grade does not exceed 24 inches near crossings.

RECOMMENDED

The depth of bioretention media ranges based on the area being treated. A 12-inch depth of bioretention media is appropriate for managing runoff from non-pollution-generating surfaces such as sidewalks while at the same time supporting plant growth. When managing runoff from pollution-generating surfaces such as streets, driveways, or vehicular loading zones, depending upon the bioretention soil media and pollutant loading targets, the depth of the media ranges from 18–24 inches.

Layering media types within the bioretention cross-section with an underdrain can help achieve water quality treatment goals. Consider using bioretention soil media for the top layer and a gravelly, sandy layer around the underdrain pipe to provide bedding for the pipe and an additional treatment layer prior to the filtered water flowing into the underdrain.

A 2- to 3-inch mulch layer may be placed over the bioretention soil media to protect plants, control weeds, lessen the need for watering, and capture many of the water pollutants before they filter through the soil.

Materials for mulching can vary. For residential streets with low volume on-street parking, chipped woody material works well within the step out zones/level areas off the curb and on the upper side slopes. Place the mulch after weeding and in the drier months of the year to allow it settle and integrate into the bioretention prior to the wet season. Compost mulch or rock mulch can be used in the bottom area and/or on the sides in the areas frequently subjected to ponding. Mulch in the bottom may only be required at installation if plants establish well or may need to be reapplied if plants in bottom of cell are pruned to the ground.

Xeriscape condition

Tall grasses

Flowers and plants for pollinators

Sedum

Schuylkill River Waterfront, **PHILADELPHIA, PA**

5 Partnerships & Performance

Policies, Programs, & Partnerships

A green infrastructure project can dramatically transform a street, adding aesthetic value and mitigating local flooding. And a citywide green infrastructure program can bring broader benefits, including climate resiliency, improved water quality, and regulatory compliance that doesn't rely solely on additional gray infrastructure investments.

Achieving these lasting benefits requires a holistic, programmatic approach to sustainable stormwater management backed by strong policy commitments, implemented with intentional, collaborative partnerships, and evaluated using a range of performance measures to document success.

Policy

Achieving success on both the programmatic and project scale depends on the development of strong policies, guidance, and tools to support green infrastructure implementation in the right-of-way. Whether motivated by federal regulatory requirements for water quality and CSOs, or by local sustainability or climate action plans, green stormwater infrastructure should be incorporated into city policies and plans at a range of scales to address stormwater management on both public and private property.

Complete Streets Policy

Many cities have addressed the inclusion of GSI practices through the adoption of Complete Streets policies that specifically require greening practices. Cities may include green infrastructure in their project development process, including as part of a Complete Streets checklist.

Green Streets Policy

Some cities have adopted specific Green Streets policies that include goals and processes for capturing and infiltrating stormwater at transportation facilities.

Local Standards & Guidance

The adoption of standard drawings, stormwater manuals and design guidelines have been a critical step in the advancement of GSI implementation in the right-of-way.

Local design guidelines provide recommendations for implementing green practices in street projects. Integrating stormwater management into project development and review, including in Complete Streets design guides or checklists, codifies goals and responsibilities for implementing green infrastructure.

Stormwater manuals have also been instrumental in compliance with regulatory and policy requirements. Design manuals can codify detailed local knowledge and best practices into a rigorous toolkit, and may include sizing and modeling criteria, preferred vegetation and media, and maintenance strategies.

Adopting **standard GSI drawings** as part of a city's standard drawing set ensures planning, design, and construction are in alignment among agencies, project goals, and construction standards.

Green Street Design Standards

Cities may take complete street design and green street priniciples a step further by developing a green infrastructure design toolkit specifically catered to local street contexts. Green street design manuals can provide detailed strategies for merging multi-modal design with ecological performance.

STREETS & PUBLIC ROW (35 - 40% OF IMPERVIOUS SURFACE)   **OPEN SPACE** (10%)   **SCHOOLS & PUBLIC FACILITIES**

The proportion of impervious surface by use and location may vary by city, but across the US, a substantial proportion of impervious surface in the city is publicly owned, whether as part of the ROW or on publicly owned land. However, many existing policy levers are focused on privately owned properties and can be used in conjunction with green street projects.

Stormwater Code/Regulations/Ordinance/Zoning

Most cities have codes, regulations, and ordinances that regulate stormwater runoff. Policies remove obstacles and create incentives for reducing stormwater runoff on both private property and public rights-of-way.

Development Fee/Incentive

Development incentives are offered to private developers who proactively use enhanced sustainable and green building and site design practices. These incentives can provide not only site-level benefits but neighborhood benefits as well, adding green features and placemaking amenities. Additionally, developers have an opportunity to realize potential tax credits, expedited permitting, or a floor area ratio bonus, depending on the structure of the incentive.

Stormwater Fee Discount/Credit

Stormwater fee discounts and credits are offered to property owners to reduce impermeable surface or use GSI practices to reduce stormwater runoff. Cities and jurisdictions that levy stormwater utility fees may offer discounts to incentivize on-site stormwater management.

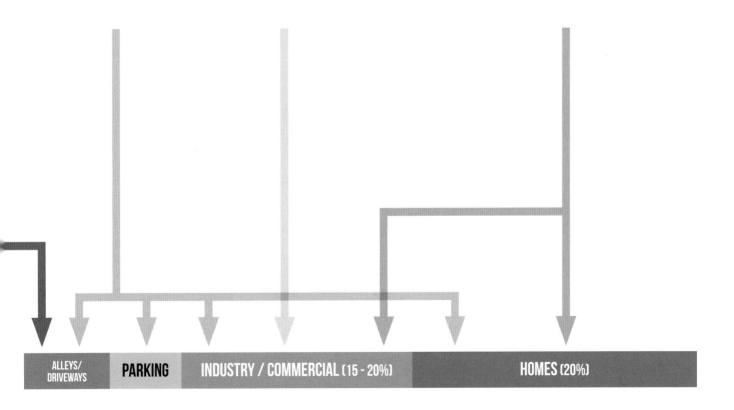

ALLEYS/DRIVEWAYS   PARKING   INDUSTRY / COMMERCIAL (15 - 20%)   HOMES (20%)

Case Study: NYC Green Infrastructure Program

Location: New York City, NY

Street Context: Dense Urban

Project Area: Citywide

Impervious Area Within Combined Sewer Tributary: 78,749 acres

Participating Agencies: New York City Departments of Transportation, Environmental Protection, Design and Construction, and Parks and Recreation

Timeline: Green Infrastructure Plan released 2010
Demonstration projects completed 2012

Cost: $1.5 billion over 20 years

GOALS

Stormwater management: Manage 1-inch rain event runoff from 10% of impervious surfaces through green infrastructure.

Cost-effective regulatory compliance: Reduce combined sewer overflow events using green infrastructure to save billions of dollars in gray infrastructure investment needs.

Water quality: Improve water quality in New York's harbor and watersheds.

Curbside bioswales, **NEW YORK, NY**

OVERVIEW

The New York City Department of Environmental Protection (DEP) is required, under a 2005 Order on Consent by the New York State Department of Environmental Conservation to reduce combined sewer overflows (CSOs), to improve water quality in and around the city. In 2012, the consent order was modified to include an ambitious green infrastructure program, anticipated to save billions over the cost of traditional gray infrastructure (sewers, holding tanks, and new or expanded wastewater treatment plants).

Street right-of-way comprises 28% of all NYC CSO drainage areas, more than any other category of City-managed land. The vast majority of this area is made of impervious asphalt and concrete, and funnels nearly all stormwater into the city's combined sewer system. Public right-of-way is the most readily accessible site type for green infrastructure and is therefore the first area addressed by DEP's program. DEP is also developing procedures for widespread implementation of green infrastructure on public and private property.

NYC developed bioswale standards and an ambitious plan of deploying an average of one bioswale per acre of the target tributary areas. Bioswales are built upstream of existing catch basins to capture runoff from the street and sidewalk, and allow for runoff from extreme rain events to continue to flow into the combined sewer system once green infrastructure is fully saturated.

The right-of-way program evolved from Greenstreets, launched by NYC Parks and Recreation decades ago to improve air quality and beautify neighborhoods with plantings in unused street rights-of-way. NYC DEP's green infrastructure explicitly aims to capture stormwater, especially in areas where traditional bioswales are impractical or inefficient.

Temporary signage on a demonstration project, **NEW YORK, NY**

KEYS TO SUCCESS

Standardized designs. An interdepartmental working group tested bioswales over a range of streets to determine design standards that could be replicated without the need for intensive site-specific design work. Final bioswale designs range from 10–20 feet in length, and 3–6 feet in width to maximize the volume of stormwater captured by each project.

Have clear siting standards. Trained staff inspect potential sites for new green infrastructure using a set list of criteria established in concert with the NYC Department of Transportation, which oversees all city rights-of-way. Due to strict criteria and the constraints of a highly built urban environment, many possible sites are initially selected and then winnowed down by disqualifying criteria (e.g., being too close to an intersection or subway line, or on ground with poor drainage qualities).

Pair green infrastructure investments with other capital projects. Every capital streets project undertaken by New York City is reviewed for potential green infrastructure inclusion. For example, New York City is building numerous Select Bus Service routes, which often require capital reconstruction of the roadway. As the street is already being reconstructed, DEP has committed to installing and maintaining bioswales in transit stops along the corridor, fast-tracking green infrastructure installation, and providing a more pleasant experience for transit riders.

Clearly communicate co-benefits of green infrastructure. Bioswales and Greenstreets improve the street's aesthetics, attract pollinators, improve air quality, and sequester carbon.

Perform community outreach while siting. The same team that sites bioswales on the street and checks for site compliance also distributes literature to residents on the street, increasing the efficiency of the team.

LESSONS LEARNED

Involve all stakeholders. Thoroughly identify all stakeholders early, including not-so-obvious actors that may be affected by new built infrastructure, such as potential upcoming real estate development, and city departments involved with roadway resurfacing.

Think about maintenance from the start. The easier green infrastructure is to maintain, the more successful it will be long-term. Choose plants that will require little maintenance, and have a plan for trash collection, clearly communicated to stakeholders, to keep bioswales litter-free and in a highly functioning state.

Clearly communicate cost savings. NYC's investment of $2.4 billion in green infrastructure is anticipated to save New York City $1.4 billion as compared to gray infrastructure substitution projects, and an additional $2 billion in deferred costs.

OUTCOMES

0.6% impervious surface runoff captured by 2015, 6% of 2030 goal.

3,830 green infrastructure assets built or under construction by 2015.

Bioswale, **NEW YORK, NY**

Curbside bioswales, **NEW YORK, NY**

Case Study: Green City Clean Waters, Green Street Program

Location: Philadelphia, PA

Street Context: Dense Urban

Project Area: CSO-based

Impervious Area Within Combined Sewer System: 28,692 acres

Participating Agencies: Philadelphia Water Department, Department of Streets, PennDOT, Southeastern Pennsylvania Transportation Authority, Commerce, Philadelphia Planning Commission

Timeline: Green City, Clean Waters *plan was published in* 2009
Implementation began in June 2011
In the first five years of the program, 111 Green Streets were constructed

Cost: $2.4 billion over 25 years

GOALS

Stormwater management: Reduce combined sewer overflow pollution by 85% and achieve water quality improvements in compliance with state and federal mandates.

Cost-effective regulatory compliance: Reduce combined sewer overflow events using green infrastructure to save billions of dollars in gray infrastructure investment needs.

Water quality: Improve water quality in Philadelphia's rivers and watersheds.

N Fairmount Avenue, **PHILADELPHIA, PA**

OVERVIEW

The Philadelphia Water Department (PWD) is required, under a 2011 Consent Order and Agreement administered by the Pennsylvania Department of Environmental Protection, to reduce combined sewer overflows and improve water quality within the city of Philadelphia. The City of Philadelphia's landmark plan to meet its goals associated with the Consent Agreement, Green City Clean Waters (GCCW), prioritizes the deployment of green infrastructure across the City of Philadelphia utilizing PWD investments, regulatory and stormwater fee modifications, and grants to manage the first inch of stormwater from over 9,500 acres of impervious surface. GCCW's GSI investments are more cost-effective than traditional gray stormwater investments, and enable the department to support investments across Philadelphia that support improvements to the City's parks, schools, and streets.

Streets represent approximately 38% of all impervious surfaces within the CSO, close to twice the second highest category of impervious surface within the CSO (residential rooftops which represent 20% of impervious surfaces within the CSO). PWD's Green Streets program utilizes the City's authority over the ROW, and ongoing multi-agency investments therein to deploy an ambitious portfolio of green infrastructure investments across Philadelphia's streets.

Philadelphia's green stormwater infrastructure investments are each designed to maximize the amount of stormwater managed at a given location. A variety of interventions, from bioswales to rain gardens, vegetated bumpouts and tree trenches are built along the ROW to capture runoff from streets and sidewalks. Aided by a Green Street Design Manual that makes the City's design standards available to consultants, engineers, and partners, PWD has been able to build projects with agency partners from PennDOT to SEPTA.

PWD's partnerships have supported other City investments. Collaborating on Safe Routes to Schools projects, Philadelphia's Water Department has been able to green bumpouts in support of reduced crossing distances near schools, increase the City's tree canopy, and invest in greening commercial corridors.

KEYS TO SUCCESS

Points of contact. Establishing green infrastructure liaisons in partner agency planning and engineering departments has been particularly helpful in facilitating the identification and creation of new Green Street partnership opportunities and resolving interagency conflicts and concerns.

Ongoing and regular communication. The creation of a Green Street Committee that meets on a quarterly basis with the Streets Department provides an open forum to explore new designs and resolve project management concerns. The Green Streets Committee is tasked with reviewing and approving pilot technologies, managing concerns related to IT and GIS systems, improving agency protocols, and discussing partnership projects.

Preview protocols. Because each Green Street in Philadelphia is designed to maximize the amount of stormwater managed at a given location, the City's Green Street program requires its partners to provide input early, and often. Project preview provided to the city's Streets Department and transit agency resolves conflicts early in the design stage, and saves money and time in the long run.

Stormwater curb extension, Queen Lane, **PHILADELPHIA, PA**

Leverage funding for complete streets. The City of Philadelphia has been able to leverage PWD's investments in green infrastructure for larger, more comprehensive grant funded projects. For instance, the Philadelphia Water Department's $3.25 million investment in a bioswale along a two-mile corridor in North Philadelphia supports an $18 million investment in the corridor mixing City and Federal grant (TIGER) dollars for an innovative streetscaping project.

Expand opportunities through partnerships. Working with City agencies (from the Parks Department to the Commerce and Streets Departments) allows PWD to share project costs and get access to both land and opportunities to develop projects that manage stormwater runoff from the right-of-way.

LESSONS LEARNED

Design is iterative. Continual monitoring and evaluation of system performance and maintenance management provides critical feedback to the Green Streets design teams. Their feedback helps PWD design more-effective, cheaper, and impactful projects moving forward.

Ongoing communication is key. Whether it is through designated point persons, or regular meetings between agency leaders, open and frequent communication at all levels (from staff to executive) is critical for programmatic success

OUTCOMES

Five years in, GCCW has implemented over 1,600 green stormwater tools on 440 sites public and private across the city, reducing pollution from sewer overflows by 1.5 billion gallons each year.

The program has leveraged $51 million of investment in streets, parks, schools, and public housing as a result of green infrastructure projects, and has created 430 new green industry jobs.

55th & Hunter Streets, **PHILADELPHIA, PA**

Green bus shelter roof, **PHILADELPHIA, PA**

Collaboration & Partnerships

Building momentum and leveraging opportunities for green infrastructure requires deliberate coordination. From identifying and securing new program funding mechanisms, to permitting, implementation, and maintenance, public agencies must collaborate between departments and levels of government, as well as with non-governmental partners. Strong collaboration helps build public support for green streets and leads to better-designed projects and programs.

Green alley programs, for example, require robust coordination and communication among public and private stakeholders, **SAN FRANCISCO, CA**

INTERAGENCY COORDINATION FOR PROGRAMMATIC SUCCESS

Coordinate among agencies responsible for water, streets, parks, and transit. Coordinating street design projects that include green infrastructure investments provides the opportunity to meet multiple concurrent goals. Transportation projects that require curb line changes in support of pedestrian safety or transit improvements are also opportunities to manage stormwater. Integrating street changes adjacent to publicly controlled land (such as parks or greenways) increases the volume of stormwater that can be managed as part of the project. Interagency coordination also increases the ability of street infrastructure investments to provide a triple-bottom-line benefit to the community at large.

Multi-agency projects require aligning not only priorities, but engineering and construction timelines as well. Coordination should be deliberate and regular throughout the process, from planning to design to implementation and maintenance.

Codify responsibilities. Memorandums of Agreement or Understanding are important steps in developing joint programs and building and managing joint projects. On a programmatic basis, MOAs and MOUs indicate executive leadership's enthusiasm and commitment for collaborative projects, an important signal to project managers. On a project level, memorandums are important for codifying maintenance and operational responsibilities, such as clarifying maintenance tasks and responsibilities between transportation, parks, and water & sewer departments.

Maintenance of infrastructure is a critical component of ensuring the longevity of green infrastructure systems. Evaluate each department's abilities and constraints to perform different maintenance functions, and clearly establish expectations for level of maintenance.

New York, NY

New York City's Departments of Environmental Protection, Transportation, and Parks & Recreation have a three-party MOU that outlines each agency's responsibilities related to the construction, maintenance, and operation of green stormwater infrastructure.

Partner for funding. Coordinated investments facilitate project funding by tapping into multiple different programmatic sources. Local agency funding for GSI can be used as a match to pursue grant funding from states, MPOs, or federal funding streams.

For example, funding dedicated to Safe Routes to School investments can be coupled with GSI investments to support bioretention in curb bulbouts around schools, shortening crossing distances for students, parents, teachers, and staff. Funding for stormwater street trees can augment economic development investments on commercial corridors. Or funding for a transit line can be leveraged alongside funding for stormwater management, creating opportunities for high-impact green transitways. Such coordinated investments make projects more cost-efficient and promote good fiscal stewardship of public dollars.

Often, partner agencies have different limitations on what they can fund, due to legislation or other requirements. Understanding funding restrictions is critical for project financing.

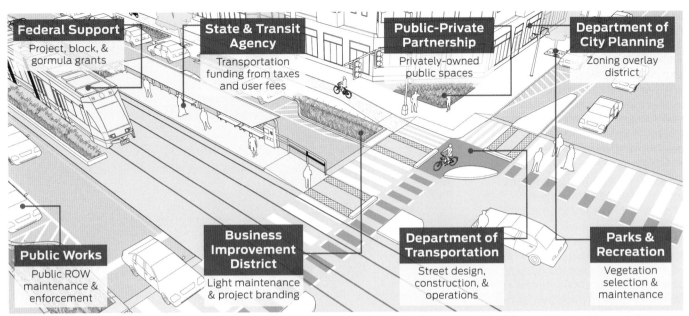

Federal Support
Project, block, & gormula grants

State & Transit Agency
Transportation funding from taxes and user fees

Public-Private Partnership
Privately-owned public spaces

Department of City Planning
Zoning overlay district

Public Works
Public ROW maintenance & enforcement

Business Improvement District
Light maintenance & project branding

Department of Transportation
Street design, construction, & operations

Parks & Recreation
Vegetation selection & maintenance

Successful green street projects are supported by a multitude of stakeholders from concept to ribbon-cutting and beyond. Above are some common examples of roles and responsibilities for designing and operating public rights-of-way.

PUBLIC–PRIVATE COLLABORATION

Regulations and procurement policies can support the development of green streets facilities. While most regulatory frameworks for infill development do not support the development of green streets directly by developers, impact fees, credits, planned unit development regulations, and green street purchasing agreements can be utilized to finance or develop green streets. Special taxing districts such as business districts and homeowners associations can also be part of a strategy to implement green infrastructure in the right-of-way.

Programs and partnerships can facilitate public-private partnerships to build green infrastructure. Whether by supporting third parties to aggregate projects along or on private property or contracting private enterprises as program managers to oversee GSI development, these partnerships may develop GSI at a lower cost. Successful partnerships carefully delineate responsibilities, facilitate monitoring, and support data and plan set sharing within contracts and partnership agreements.

Advocates are powerful allies in developing support for green street programs and garnering local project level support.

PUBLIC OUTREACH, EDUCATION & PARTNERSHIPS

When communities invest money and other resources into new programs, they often create oversight committees made up of different stakeholders and experts to help guide the public investments. When developing a program to implement bioretention in the ROW, these committees can also help with outreach and coordination to the groups that they represent.

Programs are not successful unless they implement functioning infrastructure. Many cities assist local designers, engineers, contractors and residents with technical training classes, informational materials, design standards, and manuals to assist in the successful long term implementation of the program.

Portland % For Green

In Portland, capital projects in the right-of-way contribute 1% of construction costs into the % For Green program fund. Community stakeholders can apply for grants to implement green infrastructure that treats stormwater from the public right-of-way. Funds are used for projects that go above and beyond the Stormwater Management Manual's requirements.

Non-governmental organizations can provide education and training. They can also offer a venue for volunteers to help maintain bioretention facilities, especially in residential areas or in business districts. NGOs can assist with grant applications and/or distribution of grant funding.

Cities can develop formal partnerships with community groups through grant programs or maintenance agreements, giving local residents and businesses a stake in the program's success.

San Francisco Urban Watershed Stewardship Grants

San Francisco has a competitive grant program for community-based GSI projects. Non-profits, neighborhood groups, and schools can apply for funding to plan, design, and construct green stormwater facilities.

The media can play an important role in the outreach and communication about programs and policies. Proactive communication about the purpose and benefits of the program can potentially reduce resistance and/or concerns about bioretention facilities.

Operations & Maintenance

Green stormwater infrastructure performance can improve over time if facilities are properly maintained. As vegetation establishes, roots can capture and retain more stormwater. Healthy vegetation and soil increases transpiration, reduces urban heat island effects, supports groundwater recharge, and restores natural ecological cycles and resources.

Robust and iterative operations and maintenance plans are critical to fully capitalizing on the potential of green infrastructure. Include maintenance staff in the project planning process to reduce oversights in the design and ensure that green stormwater infrastructure can achieve its full potential.

Volunteer maintenance crew, **PHILADELPHIA, PA**

ESTABLISH LEVEL-OF-SERVICE STANDARDS

Establish policy standards for an acceptable versus unacceptable level of service for maintenance of green infrastructure facilities. These performance standards must be clear and easily communicated.

Phased level-of-service standards differentiate between the establishment period and long-term maintenance of green infrastructure. During establishment, facilities may be irrigated and not receive runoff to allow plants to take root; weeding and pruning are more carefully monitored for 1–3 years after installation.

Projects located along streets with commercial or business districts may have higher standards of care for pruning and trimming compared to projects located in residential neighborhoods. It is critical during the project planning phase to review expectations for the level of service with the adjacent properties.

DEFINE ROLES & RESPONSIBILITIES

Clear responsiblities for tracking, inspection, and maintenance of green stormwater infrastructure are essential within and across departments and agencies. Establish clear roles and protocols for the care and monitoring of assets during the design process. Enact agreements or memorandums during the design process with clear delineation of responsibilities and identified partners, and clearly define the protocols for routine and corrective maintenance.

Cities under consent decree to implement green infrastructure typically enact maintenance and operations roles as part of regulatory agreements.

Maintenance may be completed by city departments or contractors as a course of implementation, with codified roles and protocols. Define clear tasks and schedules for maintenance needs. Staff training and procurement of suitable equipment are key components.

Business Improvement Districts, community organizations and stewards, and property owners may be valuable maintenance partners, provided appropriate training and education.

TRACK & MANAGE ASSETS

As green infrastructure programs scale, an accurate and current database of GSI assets is a powerful tool for maintaining a high level of system performance. Asset management may track facility locations, age, design elements (such as subsurface design), purpose of installation and regulatory reporting needs, and past inspection records.

Asset tracking systems should include information on what monitoring and maintenance is required for a specific facility. Results from monitoring can also be used to improve and adjust standard details.

Tracking GSI systems at the individual cell level, rather than at a corridor or project level, can assist with better understanding and projecting annual maintenance needs.

Create feedback loops between monitoring, maintenance, and design staff; thorough understanding of actual performance is essential to improving the design of future facilities, especially when designs are individualized for context rather than standardized for broad implementation.

SURFACE MAINTENANCE

Surface-level maintenance is necessary both for ecological performance and for aesthetic value. Remove trash, sediment, and debris from stormwater facilities and tree trenches; this routine maintenance should be completed frequently, anywhere between weekly and quarterly. Partner with a community group or Business Improvement District where possible.

Erosion repair and control, as well as structure or pavement inspection (and potentially repair), should be done at least once annually by a trained crew.

For permeable pavement systems, street sweeping and vactoring should be completed frequently, as often as weekly, to ensure that pores remain open for water to drain through. Small stones and aggregate between pavers should be replaced periodically, as small materials will gradually be removed through the course of normal wear.

Monitor bioretention facilities according to their design goals. Use water sampling monitoring or infiltration testing to ensure the facility drains as intended. Monitor the rate of sediment accumulation and adjust the frequency of cleanings and maintenance as necessary.

Identify opportunities to work with non-profit and neighborhood organizations to provide additional landscaping and cleaning, especially at high-profile locations such as a commercial district. Consider the partner organization's capacity when outlining a scope of work or maintenance contract.

VEGETATION MAINTENANCE

Watering and irrigation are essential during the establishment period of new bioretention facilities, and should be done as often as daily at first. Piped irrigation systems or watering bags can reduce the amount of labor required to nurture new plantings.

Weed removal is critical to facility health and performance, especially in the period after installation. Crews or maintenance partners may need to care for new facilities weekly.

Maintenance crew tends to vegetation and debris, **SEATTLE, WA**

Mulch and soil should be inspected periodically for loss, compaction, or other issues that can affect vegetation health.

Maintenance crews or trained partners and stewards should complete pruning or trimming typically at least once per month as part of routine superficial inspection and care. Ensure partners receive proper education so plants are not over-pruned. Perennials should also be trimmed back each winter in cold-weather climates.

At the end of a facility's design life, major maintenance or full reconstruction may be necessary to replace feature components such as soil, mulch, or vegetation.

SUBSURFACE MAINTENANCE

More involved subsurface cleaning and maintenance may be completed either semi-annually or annually, depending on what types of systems are installed, and what intensity of use is placed on the system by regional climate and performance need.

Typical subsurface tasks may include vacuum cleaning and jet-rodding of cleanouts and sewer pipes to ensure proper drainage rates and volumes.

While the expected lifespan of green infrastructure varies by cell type, construction materials, climate, and the amount of stormwater managed and treated by the facility, plan for cyclical replacement of soil and media, and the possible need for full reconstruction of stormwater facilities. These infrequent maintenance tasks may be needed after 10 to 20 years of performance, though design life varies widely and should be broadly planned for during the design process. Regular maintenance can extend the life cycle of green infrastructure assets.

Washington Avenue, **MINNEAPOLIS, MN**

Performance Measures

Stormwater solutions that make city streets into better places are built on carefully selected performance measures as much as on concrete and soil. Cities must select performance metrics that support their policy and program goals, allowing designers to keep sight of these goals throughout project development. Additionally, cities implementing green stormwater practices are often required to report on infrastructure performance and the return on investment as part of compliance reporting.

Carefully planning for strategies to measure the function and efficacy of GSI systems creates opportunities to communicate the benefits and value of sustainable stormwater infrastructure to the community.

Measure Performance for Policy Goals

Green infrastructure projects and programs should be evaluated on their contribution to city policy goals. Performance measurements are separated into three categories: ecology, mobility, and urban vitality. Primary and secondary metrics are listed below—the basic measures that are essential to understanding and communicating performance, and measures that are useful for engaging in deeper conversations to achieve near- and long-term sustainability and resilience goals.

Ecology

Green stormwater infrastructure is primarily designed to manage stormwater runoff, mitigate flooding, improve water quality, and protect local water bodies. Ecological goals vary by project and by city, and may be based on regulatory requirements or policy commitments. Ecological performance measures at the project and program scale offer guidelines for developing, sizing, and engineering a system, as well as opportunities to assess and modify designs to enhance overall performance.

Mobility

Urban stormwater streets should be designed as part of a systematic effort to make streets safer, more inviting to all users, and more accommodating of all travel modes. Measuring progress toward safety and mobility for everyone requires coordination among city departments and regional, state, and federal agencies.

Urban Vitality

Cities thrive on social activity, human connections, cultural engagement, and environmental awareness. Public health and economic opportunities are foundational to thriving cities. Operationalizing and communicating the diverse benefits of green infrastructure can open new partnership and implementation opportunities. Measures of urban vitality can be used to demonstrate efficacy and generate momentum for green stormwater programs.

PRIMARY METRICS

Water
> Volume reduction
> Water quality
> Flow rates

Soil
> Infiltration rates
> Sediment/nutrient load

Vegetation
> Plant establishment
> Tree canopy
> Species diversity

Safety
> Traffic fatalities and severe injuries
> Crash locations & characteristics
> Speed reduction

Sustainable Transportation
> Corridor mode share: number and percentage of people biking, walking, and riding transit

Access & Activity
> Parking utilization & activity
> Freight access & parking duration
> Public space canvass

Public Space
> Public space canvass
> Park density

Public Health
> Water quality
> Pollution-related illness
> Active recreation
> Obesity rate

Economy
> Property value
> Energy demand

SECONDARY METRICS

Species Habitat
> Presence of pollinators

Climate Change Mitigation
> Evapotranspiration rates
> Energy consumption

Sustainable Transportation
> Citywide mode shift: change in number and percentage of people biking, walking, and riding transit
> Transportation-related pollution, including greenhouse gas emissions

Public Health
> Heat-related illness
> Chronic illness rates
> Mental health

Economy
> Regional economic growth

Measure Performance at Program & Project Scales

Cities have deployed green stormwater infrastructure at a variety of scales, from pilot and demonstration projects to multi-million dollar program investments. Measuring and evaluating the performance of GSI systems at any scale is essential to inform design and maintenance improvements, justify future program investments, and build broad support for the co-benefits of urban stormwater streets.

Project

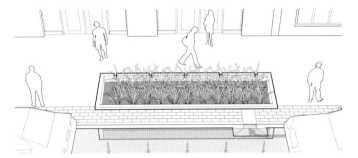

Performance measures at the project scale should evaluate how well the stormwater infrastructure meets its specific ecological and hydrological goals, how well the overall street design meets mobility goals, and how the entire project contributes to urban vitality. Project-level performance data can help support scaling up of programs in cities with or without a regulatory mandate to reduce stormwater runoff.

Project-level metrics may be tied to specific funding or regulatory requirements, especially if the project impetus is intended to resolve a specific issue or local need.

Project metrics are powerful tools for demonstrating impacts. Pilot projects should include discrete metrics to communicate benefits; strong data and evaluation is a powerful tool for building public support and leveraging future funding.

Program

Green infrastructure programs are citywide initiatives with implementation deployed at multiple sites in multiple phases. Program-scale investment and implementation are often driven by policy commitments, regulatory mandates, or zoning requirements. GSI program performance is measured at the citywide or regional scale and monitored as a connected network. Generally, network performance is required for reporting to state and federal agencies regarding regulatory requirements.

Program-level measurements are often modeled, with variables adjusted based on practice, experience, and data from existing projects. Modeling is often a less expensive metric than monitoring and is accepted by some regulators to demonstrate compliance. In cases where both modeling and monitoring data is available, models can be refined through calibration to predict performance of features under various weather trends.

Monitoring is generally expensive, and the results of monitoring can be difficult to interpret due to irregularities with a given site and specific weather events that occur during monitoring. Cities often partner with universities or other groups to perform most resource-intensive performance monitoring.

Performance Measures for Ecology

Project-Level Measures

VOLUME REDUCTION

Reducing the volume of stormwater runoff that enters the gray infrastructure system and local water bodies is a primary goal of green infrastructure projects. Volume reduction can be achieved through a combination of retention, evapotranspiration, and infiltration, depending on design goals.

Calculate the retention basin storage volume required to attenuate peak stormwater flows. Use rain gauges to collect rainfall data that sets a baseline on which to measure evapotranspiration rates and runoff. Measure stormwater runoff in the drainage area before project construction, and then monitor either outfalls or bioretention cells after implementation to evaluate runoff volume reductions.

Monitoring assets, **NEW YORK, NY**

WATER QUALITY

Whether water is being detained and slowly released into the sewer system or infiltrated into the ground, removing pollutants and improving water quality is a central goal of GSI projects for environmental restoration, public health, and regulatory compliance.

Measure the pollutant level in runoff entering the sewer system or local water bodies adjacent to the project site, both before and after project implementation, to assess improvements to water quality.

FLOOD CONTROL

Green stormwater infrastructure projects are often implemented to reduce flooding, both on streets and in neighborhoods with high water tables and significant basement flooding.

Measure volume reduction at project outfalls to evaluate the impact on flood control. Track reports of traffic closures due to street flooding, as well as 311 calls or insurance claims of basement floods in buildings adjacent to the project area.

Portland, OR

On SW 12th Street & Montgomery Street, four on-line planters have successfully retained 72–77% of runoff volume during storm flow tests. Designers also tested drawdown time, soil pollutants, and ponding depths for all four cells between 2007 and 2010.

CHANNEL PROTECTION

GSI projects are often implemented to mitigate degradation of downstream rivers and creek beds. Urban stormwater runoff—especially frequent, high-volume flows—can cause riverbank erosion, which in turn can cause severe damage to bridge and sewer infrastructure or waterfront property. Channel degradation also negatively impacts wildlife habitat and biodiversity.

Direct measurements of the riverbanks and creek beds will give an indication of the project's impact on downstream channel protection.

DRAWDOWN TIME

Bioretention facilities capture and store water before slowly releasing it into the sewer system or infiltrating it into the ground. Captured water must drain within the specified drawdown time, generally 24 to 72 hours, to limit the duration of ponding and ensure facilities can accommodate the next storm.

In some climates, drawdown time is especially important to prevent mosquito larva from developing in ponded water.

INFILTRATION RATES

Stormwater facilities, permeable pavements, and soil media are evaluated on their ability to infiltrate water. Infiltration rate, described in inches per hour, measures the rate at which water passes through the surface media, such as exposed soil or porous pavement.

Infiltration rates vary depending on project location, soil or pavement porosity, the soil's ability to treat pollution and sediment, and the project's distance from the water table. Other factors determining infiltration rates include seasonal changes to groundwater levels, shallow bedrock or other limiting soil layers, and historic industrial land use at the site with residual pollutant loads.

Infiltration rates should be designed to reduce pollutant loads passing through the soil media to the water table. Test the infiltration rate at the site before and after project installation to measure success. Continue to test for changes in infiltration rate over time.

POLLUTANT LOADS

Rain that falls on the ground collects sediments, nutrients, and heavy metals as it flows across the street surface and into the stormwater management facility. Capturing and filtering these pollutants at the stormwater cell prevents downstream pollution or algae blooms that pose public health risks and degrade the environment.

Measure suspended solids in stormwater runoff as a proxy for measuring nutrient and heavy metal loading. Requirements for measuring water quality and pollutant loads are often determined by a consent order.

PLANT HEALTH

The ability of plants to survive and thrive is crucial to the green infrastructure's stormwater management performance. Maintaining healthy plants is also key to community buy-in.

Plant health depends on soil mix, pollutant load, runoff flow intensity, salt spray, risk of trampling, and hardiness against flooding or drought. Species diversity is an important factor in providing more robust ecosystem services.

Evaluate whether plants are able to establish in the stormwater cell. Monitor plant health as a metric of project success.

SPECIES HABITAT

Native bee populations have declined in recent years due to habitat loss and pesticide use, posing a grave risk to the food crops that rely on pollination. In urban areas, human development threatens habitat for many pollinator species, including butterflies, moths, beetles, and hummingbirds.

Vegetated stormwater infrastructure with diverse, native plants can increase habitat for bees and other pollinators, restoring important ecological connections. Habitat restoration in urban areas is a foundational step toward preserving healthy bee colonies, which support food crops for human consumption.

Monitor the presence of pollinators at green infrastructure projects.

Austin, TX

Austin City Council passed a 2015 resolution to incorporate native milkweed into city-owned properties to increase and improve pollinator habitat. Since then, the Austin Watershed Protection Department has included milkweed plants in green stormwater infrastructure projects to attract bees, monarch butterflies, and other pollinators.[1]

Performance Measures for Ecology continued

Program-Level Measures

GREEN ACREAGE

Green stormwater infrastructure programs are implemented to reduce impervious surface cover at a citywide scale, reducing barriers to the natural hydrological cycle. Many cities have adopted Green Acreage goals—the total land area where the first inch of rainfall that falls on impervious surface during each storm event is managed by green infrastructure.

Measure the reduction in percentage of impervious surface land area over the course of program implementation to evaluate and demonstrate progress toward this goal. Track and communicate progress about the increase in green acreage with the community to build broad citywide support.

Geographic Information System (GIS) tools can offer powerful and cost-effective methods to model and estimate vegetated cover on a large scale.

Philadelphia, PA

The Philadelphia Water Department's Green City, Clean Waters program tracks greened acreage toward meeting their consent order goals. The City added over 1,100 green stormwater management facilities between 2011 and 2016.[2]

WATERSHED HEALTH

At a programmatic level, green stormwater infrastructure is designed to protect and improve watershed health. Watersheds transcend jurisdictional boundaries, and healthy watersheds are critical for public health, safe drinking water, and wildlife habitat.

Integrated assessments of watershed health include landscape condition, hydrology, geomorphology, habitat, water quality, biodiversity, and vulnerability to climate change. Learn more from the EPA's Healthy Watersheds program.[3]

TREE CANOPY

Many cities have specific goals to increase citywide tree canopy coverage. Calculate the increase in tree canopy coverage as part of a comprehensive GSI program.

Green infrastructure projects and increased tree canopy improve local air quality by removing pollutants and intercepting particulate matter. Measure local air quality progressively as greening programs are scaled.

New York, NY

NYC Parks & Recreation maintains a citywide tree map with every street tree geolocated and measured. Tree-level data includes species, trunk diameter, and estimates for contributions to stormwater infiltration, evapotranspiration, and energy demand savings, both as a direct measure and with an estimate for economic value.[4]

REGULATORY COMPLIANCE

In cities with a combined sewer system, heavy storms can overwhelm the water treatment plant capacity and excess runoff is discharged into nearby water bodies, often causing serious water pollution and posing public health risks. The Clean Water Act regulates the discharge of pollutants into water bodies. Many cities are under consent decrees with the EPA to reduce water pollution or combined sewer overflows, and many cities have local stormwater management water quality requirements in addition to federal regulations.

Green stormwater infrastructure is part of a citywide strategy to mitigate combined sewer overflow events and comply with federal water quality requirements. Measure the reduction in frequency and severity of combined sewer overflow events as a performance measure of GSI programs.

New York, NY

The Department of Environmental Protection tracks GSI assets and monitors progress toward its consent order goals using a public mapping tool that displays construction progress and total implementation by combined sewer overflow tributary area.[5]

FLOOD CONTROL

Many cities across the country have experienced record-setting rain events in the last ten years. Cities are experiencing 100- and 500-year storm events with more frequency, straining city storm sewer systems. These events have the ability to significantly impact the operation of streets and low-lying areas. Extreme rain events strain local wastewater collection systems, resulting in overflows and system back-ups.

Although GSI systems are typically not designed to mitigate extreme rain events, they can provide short-term, interim storage capacity for lesser events that reduce the cumulative impact during extreme rain events. GSI projects can also reduce local flooding on streets and in neighborhoods with high water tables and significant basement flooding.

Measure volume reduction at project outfalls to evaluate the impact on flood control. Track reports of traffic closures due to street flooding, as well as 311 calls or insurance claims of basement floods in buildings adjacent to project areas.

Minneapolis, MN

Minneapolis Public Schools, in partnership with the Holland Neighborhood Improvement Association and the Mississippi Watershed Management Organization, collaborated to develop the state's first green campus complete with permeable pavement, tree trenches along the right-of-way, and stormwater storage tanks to mitigate extreme flooding in the neighborhood. The project is projected to capture and treat approximately 1.5 million gallons of runoff.[6]

CLIMATE CHANGE MITIGATION

At a program level, green infrastructure provides shade and creates a cooling effect, reducing building energy consumption and the related greenhouse gas emissions.

Capturing more stormwater at the source reduces the electricity demand and the related greenhouse gas emissions at water treatment facilities.

When implemented at a citywide scale, green stormwater infrastructure is an important part of creating a livable city where people feel safe and comfortable walking. Shifting single-occupancy vehicle trips to walking and bicycling reduces greenhouse gas emissions.

Measure citywide greenhouse gas emissions over time and assess the role of green infrastructure in climate change mitigation.

Performance Measures for Mobility

Project-Level Measures

TRAFFIC SAFETY

Safety is a fundamental responsibility of city government. Streets should be designed to allow safe mobility for people walking, biking, taking transit, driving, or enjoying public space.

Stormwater streets, like all streets, should be evaluated for their traffic safety results. Green stormwater infrastructure should incorporate, and be incorporated into, street designs that advance Vision Zero goals to eliminate traffic fatalities and serious injuries. Evaluating safety is important in both building the case for future projects and evaluating the design details of existing projects.

In the design process, analyze crash counts by location and crash type (e.g., rear-end collisions, turning-vehicle collisions, crashes between modes). Work with police departments to simplify data collection and reporting as needed. Crash locations should be geocoded as accurately as possible to understand locations where crashes are more frequent. In evaluating projects, it may be difficult to determine whether crashes are related to green infrastructure. However, crash patterns at a location indicae that further design attention may be necessary.

An important, discrete metric to study is the number of people killed or severely injured (KSI) on city streets. At a corridor level, collect data about fatalities and injuries over a multi-year period (~3 years) to gather an adequate sample size to understand risk conditions before and after project implementation.

Speeding is one of the key issues related to unsafe streets, and a leading indicator of a design's safety results. Green infrastructure should be implemented as a traffic calming strategy, creating safe places for people while reducing motor vehicle speeds. Evaluate whether street redesign projects have reduced the number and frequency of speeding vehicles. 85th percentile speed is commonly used, but 95th percentile speed is a more accurate measure of the most dangerous speeding behavior.

MOBILITY

On a project level, green infrastructure projects can be implemented alongside street improvements that support active transportation, such as transit boarding bulbs or protected bikeways. These investments enhance biking and walking access or transit service quality, and may improve or maintain overall traffic flow by organizing street operations and reducing turn conflicts.

Average transit travel time through project segments can be important to identify benefits from transit-supportive street design projects. Stop-level boarding data and stop delay time are necessary to evaluate the granular benefits of more-efficient transit boarding and in-lane stop configurations. Analyze changes in motor vehicle travel speed to build a complete picture of mobility outcomes.

Shade trees and vegetated planters improve the experience of people waiting for a bus or train, providing a more hospitable human environment at transit stops. Conduct rider surveys to collect qualitative data about experience at transit stops before and after GSI implementation.

SUSTAINABLE TRANSPORTATION

Integrating GSI into streets can be part of a comprehensive strategy to reallocate unused or underused street space to sustainable travel modes: walking, biking, and transit.

At a project level, collect and analyze mode share, or the number and percentage of people on a corridor traveling by each mode, especially for projects that include pedestrian, biking, and transit infrastructure improvements. Track changes to each mode separately as well, indicating whether a large increase has occurred in a newly served mode. Successful GSI projects enhance the public realm and encourage sustainable travel mode choices.

ACCESS & ACTIVITY

For many projects, success hinges on maintaining access to local destinations and businesses, and changes to on-street parking are often part of the discussion. Measure parking utilization to determine where parking demand is low and curbside space is easiest to repurpose for green infrastructure.

When reallocating street parking for bioretention, calculate parking spaces not just on the project corridor but also within 1 to 2 blocks, located on- and off-street, that are available for destination access. For instance, 50% parking loss on a project street may only constitute 1–2% of total parking availability within a block. Distinguish between changes in high-value loading zones, moderate-value metered parking, and lower-value storage parking.

Conduct public space canvasses to identify user patterns of streetscape improvement projects, and identify successful strategies for increasing social activity.

San Francisco, CA

The sustainable streetscape project on Newcomb Avenue is not only reducing peak stormwater flow rates by 72–83%, but has reduced 85th percentile speed 3 mph in each direction and halved the volume of motor vehicle traffic cutting through this neighborhood street.[7]

Program-Level Measures

//

SUSTAINABLE TRANSPORTATION

At a citywide level, evaluate the GSI program's contribution to overall safety and mode split policy goals.

Compare year-to-year changes in citywide biking, walking, and transit mode share to observe long-term trends as green street improvement programs are brought to scale. Use intercept survey data to collect a more qualitative understanding of why people choose to bike, walk, or take transit rather than drive, and include questions about how green infrastructure affects their mobility experience.

Air quality and pollution reduction are essential to cultivating inviting streets for walking and bicycling. If people are driving less and walking more because of an enhanced streetscape condition, CO_2 reductions can be realized. Document changes in mode share to demonstrate progress towards a more sustainable urban design and transportation system.

At the program scale, atmospheric CO_2 can be measured through the reduction of energy use, whether through lowered energy demand due to shading, or reduced energy demand from water treatment facilities, thereby reducing CO_2 emissions from power plants.

Transportation-related pollution can be estimated at a program level using per mile emissions estimates multiplied by total vehicle miles traveled (VMT) at corridor, neighborhood/area, or citywide scales. As VMT is reduced and trips are shifted to more efficient modes, green infrastructure must mitigate less air pollution.

Performance Measures for Urban Vitality

Project-Level Measures

PUBLIC SPACE & STREET LIFE

Accessibility, connectivity, and vibrancy are requirements for a well-functioning urban environment. GSI projects can contribute to the creation of a cohesive and legible streetscape and safe, beautiful public spaces.

Conduct intercept surveys of people walking, bicycling, and sitting in public spaces to gather qualitative data about their perceptions and experiences of safety, happiness, stress, and well-being. Use metrics like the gender ratio of bicyclists or the ratio of children and seniors to young adults to evaluate comfort for all ages and abilities.

Cities are thirsting for public space. Measure acreage or square feet of space reclaimed from car-only street space. Even hardscape, if relevant for public life and activity, is valuable square footage.

Measure public life before and after project implementation and use before/after photography to document changes. To evaluate projects systemactically, compare project data with comparison corridors for a control case. Before/after data and photographs, as well as comparative studies, are critical tools for project evaluation and case-making for future projects.

Measure human activity in public spaces through public life canvassing. Count people using the street for stationary activities in a series of walk-through "snapshots" of a space.

ECONOMIC BENEFITS

Green infrastructure investments provide economic benefits to surrounding households, businesses, and communities.

Though GSI projects can improve adjacent property values, it can be difficult to collect adequate samples of data to be more than anecdotal. Include community members and residents throughout the design process to ensure GSI projects meet community needs and are not perceived as a tool for displacement, especially among renters who may not accrue the benefits of increased property values.

Use metrics like retail and residential occupancy/vacancy rates and sales tax revenue for local businesses along the project corridor. Conduct intercept surveys of customers visiting businesses to demonstrate approximate value of purchases made or the number of potential customers visiting the store.[8]

Green infrastructure projects and programs create stable, localized jobs ranging from construction to landscaping and maintenance to monitoring and performance management. Count direct green industry jobs created by city projects or programs, and estimate indirect and induced jobs created by program expansion.

GSI projects, especially those with shade trees, can reduce building energy demand.[9] Measure energy consumption before and after project implementation, and continue to measure over time as trees mature and provide more shade.

Chicago, IL

The City of Chicago's Greencorps is a nine-month program that offers full-time employment, job training and skill development, and professional networking opportunities for residents who face barriers to employment. Greencorps members assist with green infrastructure maintenance while preparing for green industry careers. In the first 16 years of the program, 75% of program graduates proceeded to employment or advanced positions with Greencorps.[10]

Program-Level Measures

HEAT ISLAND EFFECT

Urban heat islands, which can be significantly warmer than surrounding rural areas, are linked with an increase in heat-related illnesses. The urban heat island effect also drives an increase in energy demand for air conditioning, thereby increasing energy bills and energy-related greenhouse gas emissions.[11] The urban heat island effect occurs in dense areas with high percentages of land cover that absorb or hold heat, such as pavement, buildings, or other hardscape.

Temperature data is not a direct indicator of GSI performance, but can provide clues to inform future GSI system designs to better reduce local temperatures through evapotranspiration and shading.

Measuring reductions in the urban heat island effect requires a combination of techniques, including measuring the energy reflected and emitted from various surfaces in cities, as well as air temperature measurements.

Phoenix, AZ

Phoenix, which has a dramatic urban heat island effect resulting in city temperatures that are 15 degrees warmer than adjacent desert and farmland, charted an ambitious roadmap to plant urban forests and street trees. Phoenix aims to achieve an average 25% shade canopy coverage for the entire city by 2030 as part of a strategy to address the urban heat island effect, decrease energy consumption and costs, reduce stormwater runoff, and contribute to a more pleasant public realm.[12]

PUBLIC HEALTH

Greening in urban environments has been attributed to improving mental, physical, and social well-being.

Water quality improvement is usually the most significant and directly observable public health benefit from green infrastructure programs. Measuring reductions in combined sewer outfall events, as well as the presence of pollutants, harmful bacteria, and metals in regional surface and groundwater are basic components for monitoring and communicating program efficacy.

Air quality is also readily measurable in the urban environment; trees and vegetation absorb air pollutants and intercept particulate matter from the atmosphere that cause chronic diseases such as asthma. Measure indicators in the atmosphere such as nitrogen dioxide and street-level particulate matter.

Cities can work with institutions or public health researchers to understand the diverse human impacts of greened streets. Monitoring changes to asthma rates and heat-related illnesses among nearby residents may indicate changes to the physical health impacts of streets.

Additionally, as streets become more inviting to active use, health indicators such as obesity and cardiovascular disease can verify the benefits of more active urban environments.

Finally, mental health can be improved by urban green space; proximity and access to urban green spaces reduce stress and anxiety.[13] Survey street users to understand how they perceive greened versus unimproved streets, documenting how they react to or perceive stressful events, with or without green spaces.

EQUITY

An equitable green infrastructure program distributes green space among neighborhoods with a conscious recognition of the need to avoid harm to vulnerable people, as low-income neighborhoods have historically borne a greater risk for exposure to pollution and toxins.[14] Any negative impacts, such as flooding, must not be disproportionately borne by one community.[15]

Analyze the spatial distribution of GSI facilities and tree canopy and assess whether resources are being invested equitably. Use flooding models to supplement information from the public (such as 311 calls or constituent complaints) to inform and select project locations, and ensure that projects are being implemented across neighborhoods regardless of socio-economic factors.

Portland, OR

Portland's Bureau of Environmental Services completed a report in 2010 documenting the impacts of its Gray to Green program on residents using a wide variety of social and economic indicators. Researchers found broad improvements to resident satisfaction, economic value, and air and water quality around green project locations.[16]

6 Resources

Glossary

Biofiltration is the process of removing particulate matter and other pollutants by filtering stormwater runoff using biological material to detain and degrade pollutants. Biofiltration is a technique used in stormwater management that uses living plant material to process stormwater runoff.

Biofiltration planter refers to a facility designed to capture and treat stormwater runoff, removing pollutants using plant material to infiltrate and support settlement of particulate matter. Biofiltration planters can be designed with a permeable or impermeable base to support the conveyance of treated runoff after it is treated using biological methods.

Bioretention refers to the process of capturing stormwater runoff, absorbing and retaining pollutants, and then infiltrating, transpiring, or evaporating the water.

Bioretention facilities are treatment areas, often designed as shallow landscaped depressions, that function to capture and manage sediment and stormwater runoff. They are designed with a soil mix and plants adapted to the local climate and receive stormwater from a contributing area, such as the street. Bioretention facilities can be designed to reduce stormwater runoff quantity and surface flow rates and remove or reduce sediment and pollutants from stormwater runoff.

> » Stormwater facilities
> » Bioswale
> » Rain garden

**Note that in many cities, these terms are used interchangeably. In some cities, these terms have more precise definitions.

Bioretention swale is a shallow bioretention facility with sloped sides on all sides of the facility, designed to capture, treat, and manage stormwater runoff from a contributing area.

> » Rain garden
> » Bioretention basin

Bioretention planter refers to a bioretention facility with walled vertical sides on all four sides of the facility, a flat bottom area, and a large surface capacity to capture, treat, and manage stormwater runoff from a contributing area.

> » Stormwater planter
> » Planter boxes
> » Bioretention cell
> » Raised planter
> » Roadside planter

Bioretention facility with underdrain, regardless of type, is a cell constructed with an underdrain pipe to collect the water that has filtered downward through the bioretention media layer(s).

Combined sewer system is a sewer system that collects stormwater runoff in pipes that also convey wastewater and sanitary sewer flows.

Combined sewer overflow is an overflow event that occurs in combined sewer systems when the volume of stormwater and wastewater exceeds the capacity of the sewer system or treatment plant, often due to a storm event. When this occurs, untreated wastewater and stormwater discharge directly into receiving water bodies, such as rivers and lakes.

Curb extension, also sometimes referred to as curb bulb or bump-out, is an expansion of the curb line into a portion of the roadway either for a portion of a block, at a corner, at an intersection, or midblock. At the curb extension, the curb-to-curb roadway width is reduced. Curb extensions can enhance pedestrian safety by reducing crossing distances and provide additional space for sidewalk amenities, bioretention facilities, transit boarding, and/or street trees.

> » Bump-out
> » Curb bulb
> » Neckdown

Detention refers to the practice of collecting and holding stormwater runoff temporarily in surface or subsurface storage facilities before slowly releasing the water at a controlled rate, preventing flooding and erosion. Systems designed for stormwater detention can be vegetated or not.

Dispersion is the release of stormwater runoff such that the flow spreads over a wide area for treatment and infiltration.

Definition	Other Terminology Used

Evapotranspiration is the process by which water is evaporated through soil and plant surfaces and transpired through plant leaves.

Green gutter is a narrow and shallow landscaped strip along a street's curb line. It is designed to manage stormwater runoff by placing the top of the planting media in the green gutter lower than the street's gutter elevation, allowing stormwater runoff collected in the street's gutter to flow directly into the green gutter.

See Philadelphia's *Green Streets Design Manual* or Denver's *Ultra Urban Green Infrastructure Guidelines* for more information.

Green stormwater infrastructure (GSI) encompasses techniques used to collect, filter, and manage stormwater runoff from streets, sidewalks, parking lots and other impervious surfaces, and direct the runoff to engineered facilities that use natural processes to treat and manage the water. Examples of green stormwater infrastructure include bioretention facilities, stormwater trees, and permeable pavements.

» Green infrastructure
» Sustainable stormwater infrastructure

Hybrid bioretention planter is a bioretention facility with a combination of a vertical wall and graded side slopes. In the right-of-way, the wall side may be on either the sidewalk side or street side with the remaining three sides of the bioretention facility having graded side slopes.

Impervious refers to a material or layer that is designed and constructed to prevent fluid from passing through.

» Impermeable

Impervious surface is a non-vegetated surface area which prevents the entry of water into the soil, causing water to run off the surface in greater quantities or at an increased rate of flow than would occur under natural conditions prior to development. Common impervious surfaces include, but are not limited to, paved streets, sidewalks, and parking lots; asphalt and concrete paving; gravel roads; and other surface treatments that impede natural infiltration of stormwater.

» Hardscape

Infiltration is the process of absorbing stormwater through the ground surface and into the soil below.

Infiltration rate is the rate at which water enters the soil, usually measured in millimeters of water per hour.

Inlet is an opening in the curb or a pathway to convey stormwater from the surrounding street catchment area into the bioretention facility.

Municipal Separate Sewer System (MS4) is a sewer system that collects and conveys stormwater in pipes separate from sanitary sewer lines.

Permeable refers to anything permitting fluid to pass through.

» Pervious

Permeable pavement refers to either pervious concrete, porous asphalt, permeable interlocking concrete pavers, or other forms of pervious or porous paving material intended to allow passage of water through the pavement section. See ASCE's *Permeable Pavements Recommended Design Guidelines* for more information.

Planting strip is the area between the curb and the sidewalk that is landscaped with plants, grass, and/or street trees.

» Boulevard
» Furnishing zone
» Sidewalk buffer
» Tree lawn

Definition	Other Terminology Used
Presettling zone is an area designed to passively collect suspended debris, particulate matter, and sediment at the upstream end of a stormwater facility.	» Sediment trap » Forebay » Sediment collection pad
Retention is the practice of capturing and holding stormwater on-site to reduce runoff to sewer systems. Water is then evaporated, transpired through plants, or infiltrated through the soil.	
Stormwater runoff is water from rain or melting snow that "runs off" hardscape such as asphalt and concrete. Areas with higher proportions of impervious surface cover, such as streets, parking lots, and buildings, generate more stormwater runoff because rain is not able to infiltrate into the ground as it would in pre-development conditions.	» Runoff
Stormwater tree is a tree planted in a tree well or tree pit, designed to maximize stormwater retention. The system can be designed to have walled sides, subsurface cells, structural soil, or be depressed below grade to retain stormwater. The soil media is designed to easily infiltrate stormwater and is typically sited below the street's gutter elevation, allowing the tree to manage stormwater runoff from the street or sidewalk. Stormwater infiltrates through the soil, transpires through the tree's leaves, or drains to a connection in the storm sewer network.	» Tree well » Tree pit » Tree box
Stormwater tree trench is a subsurface trench that connects a linear series of street tree wells, distributing stormwater flow among a series of trees. The trees share the same soil volume free of significant subsurface disruptions.	» Tree trench

Endnotes

Streets as Ecosystems

STREETS ARE ECOSYSTEMS

1   Marla C. Maniquiz-Redillas and Kim Lee-Hyung, "Evaluation of the capability of low-impact development practices for the removal of heavy metal from urban stormwater runoff," *Environmental Technology* (2016): 1–8.

2   American Society of Civil Engineers, *2013 Report Card for America's Infrastructure: Wastewater*, accessed November 2016, www.infrastructurereportcard.org.

3   City of New York Department of Environmental Protection, *NYC Green Infrastructure Plan: A Sustainable Strategy for Clean Waterways* (New York: City of New York, 2010), accessed November 2016, http://www.nyc.gov/html/dep /pdf/green_infrastructure/NYCGreenInfrastructurePlan _LowRes.pdf.

4   NOAA National Centers for Environmental Information, *Billion- Dollar Weather and Climate Disasters: Overview*, accessed November 2016, https://www.ncdc.noaa.gov /billions/overview.

5   United States Environmental Protection Agency, "National Pollutant Discharge Elimination System (NPDES): Combined Sewer Overflows (CSOs)," accessed Nov. 2016, https://www .epa.gov/npdes/ combined-sewer-overflows-csos.

6   Harriet Festing et al., *A RainReady Nation: Protecting American Homes and Businesses in a Changing Climate* (Center for Neighborhood Technology, 2014).

7   Claudia Tebaldi, Benjamin H. Strauss, and Chris E. Zervas, "Modelling sea level rise impacts on storm surges along US coasts," *Environmental Research Letters* 7, no. 1 (2012): 014032.

8   David J. Nowak and Eric J. Greenfield, "Tree and impervious cover change in U.S.," *Urban Forestry & Urban Greening* 11, no. 1 (2012): 21–30.

9   Wen Liu, Chen Weiping, and Peng Chi, "Assessing the effectiveness of green infrastructures on urban flooding reduction: A community scale study," *Ecological Modelling* 291 (2014): 6–14.

10 Noelwah R. Netusil, Zachary Levin, Vivek Shandas, and Ted Hart, "Valuing green infrastructure in Portland, Oregon," *Landscape and Urban Planning* 124 (2014): 14–21.

Planning For Stormwater

1   United States Environmental Protection Agency, "Policy Memos," accessed November 2016, https://www.epa.gov /green-infrastructure/policy-memos.

2   City of Los Angeles Department of Buildings and Safety, *Guidelines for Storm Water Infiltration,* (Los Angeles, CA: City of Los Angeles, 2014), accessed November 2016, http://www .lastormwater. org/wp-content/files_mf/appxhfinal.pdf.

3   Los Angeles Bureau of Sanitation, California State Polytechnic University Department of Landscape Architecture, and UCLA Extension Department of Landscape Architecture, *Greenways to Rivers Arterial Stormwater System (GRASS) Phase 1: Summary Report*, http://www.lastormwater .org/wp-content/files_mf/grasssummaryreport.pdf.

Stormwater Streets

1   Kendra K. Levine, *Curb Radius and Injury Severity at Intersections*, (Berkeley, CA: Institutue of Transportation Studies Library, 2012).

2   Yingling Fan, Andrew Guthrie, and David Levinson, "Perception of Waiting Time at Transit Stops and Stations" (working paper, Center for Transportation Studies, University of Minnesota, Minneapolis, 2016).

Stormwater Design Elements

GREEN STORMWATER ELEMENTS

1   United States Access Board, "Chapter 4: Accessible Routes," (Washington, DC: US Department of Justice), accessed November 2016, https://www.access-board.gov/guidelines -and-standards/buildings-and-sites/about-the-ada-stand ards/ ada-standards/chapter-4-accessible-routes.

2   United States Access Board, "Chapter 4: Accessible Routes," (Washington, DC: US Department of Justice), accessed November 2016, https://www.access-board.gov/guidelines -and-standards/buildings-and-sites/about-the-ada-stand ards/ ada-standards/chapter-4-accessible-routes.

3   Geoffrey H. Donovan and David T. Butry, "Trees in the city: Valuing street trees in Portland, Oregon," *Landscape and Urban Planning* 94, no. 2 (2010): 77–83.

4   Massachusetts Department of Transportation, "Chapter 3: General Design Considerations," in *Separated Bike Lane Planning & Design Guide* (Boston: MassDOT, 2015).

5   United States Environmental Protection Agency, "Soak Up the Rain: Trees Help Reduce Runoff," accessed Nov. 2016, https://www.epa.gov/soakuptherain/ soak-rain-trees -help-reduce-runoff.

6   United States Access Board, "Chapter 3: Building Blocks," (Washington, DC: US Department of Justice), accessed November 2016, https://www.access-board.gov/guidelines -and-standards/buildings-and-sites/about-the-ada- standards/ ada-standards/chapter-3-building-blocks.

7   See ASCE guide for recommendations on ratio of run-on to permeable pavements, ASCE *Permeable Pavements Recommended Design Guidelines*.

GREEN INFRASTRUCTURE CONFIGURATIONS

1   Marina Lagune-Reutler, Andrew Guthrie, Yingling Fan, and David M. Levinson, "Transit Riders' Perception of Waiting Time and Stops' Surrounding Environments," *Transportation Research Board*, 2016, Retrieved from the University of Minnesota Digital Conservancy, http://hdl. handle.net/11299 /180075.

2   United States Access Board. "Chapter 4: Accessible Routes," (Washington, DC: US Department of Justice), accessed November 2015, https://www.access-board.gov/guidelines -and-standards/buildings-and-sites/about-the-ada-stand ards/ ada-standards/chapter-4-accessible-routes.

BIORETENTION DESIGN CONSIDERATIONS

1   American Mosquito Control Association, "Life Cycle," accessed November 2016, http://www.mosquito.org/life -cycle.

2   Mark Maimone, Daniel E. O'Rourke, James O. Knighton, and Casey P. Thomas, "Potential Impacts of Extensive Stormwater Infiltration in Philadelphia," *Environmental Engineer* 14 (2011): 29–39.

Partnerships & Performance

PERFORMANCE MEASURES

1   City of Austin, "What's the Royal Buzz about Pollinators?" accessed December 2016, http://www.austintexas.gov/blog /what%E2%80%99s-royal-buzz-about-pollinators.

2   Philadelphia Water Department, "Green City, Clean Waters," accessed November 2016, http://www.phillywatersheds. org /what_were_doing/documents_and_data/ cso_long_term _control_plan.

3   United States Environmental Protection Agency, "Healthy Watersheds: Protecting Aquatic Systems through Landscape Approaches," accessed November 2016, https://www.epa .gov/hwp.

4   New York City Department of Parks and Recreation, "New York City Street Tree Map," accessed November 2016, https:// tree-map.nycgovparks.org/.

5   New York City Department of Environmental Protection, "DEP Green Infrastructure Program Map," accessed December 2016, https://www.arcgis.com/home/webmap/viewer.html ?webmap=0061d39df78d41978b9a662fb8d17981.

6   Mississippi Watershed Management Organization, "News Release: Edison High School Leads the State in Going Green," August 16, 2016, accessed December 2016, http://edison .mpls.k12.mn.us/uploads/edison-green-campus-news-re lease-2016-08-16.pdf.

7   San Francisco Public Utilities Commission, "Newcomb Avenue Green Street Monitoring Report Rainy Seasons 2011- 2012 and 2012-2013," accessed December 2016, http:// sfwater. org/modules/showdocument.aspx?documentid =8301.

8   Noelwah R. Netusil, Zachary Levin, Vivek Shandas, and Ted Hart, "Valuing green infrastructure in Portland, Oregon," *Landscape and Urban Planning* 124 (2014): 14–21.

9   Hashemk Akbari, "Energy Saving Potentials and Air Quality Benefits of Urban Heat Island Mitigation," Lawrence Berkeley National Laboratory (2005), accessed November 2016, http:// www.osti.gov/scitech/biblio/860475.

10   Institute for Sustainable Communities, *Greencorps Chicago: Program to Reintegrate Ex-Offenders Into the Workforce* (Chicago, IL, 2010), http://www.sustainablecommunities leadershipacademy.org/resource_files/documents/Chicago ,%20IL_1.pdf.

11 United States Environmental Protection Agency, "Heat Island Effect," accessed November 2016, https://www.epa.gov /heat-islands.

12 Lyssa Hall, City of Phoenix Tree and Shade Management Task Force, City of Phoenix Parks and Recreation Department, and James Ritter, *Tree and Shade Master Plan* (Phoenix, AZ: City of Phoenix, 2010), accessed December 2016, https://www .phoenix.gov/parkssite/Documents/071957.pdf.

13 Ian Alcock et al., "Longitudinal effects on mental health of moving to greener and less green urban areas," *Environmental Science & Technology* 48, no. 2 (2014): 1247–1255.
- - - - - - - - - - - - - -
Konstantinos Tzoulas et al., "Promoting ecosystem and human health in urban areas using Green Infrastructure: A literature review." *Landscape and Urban Planning* 81, no. 3 (2007): 167–178.

14 Amy Vanderwarker, "Water and Environmental Justice," in *A Twenty-First Century US Water Policy* (Oxford, England: Oxford University Press, 2012), 52.

15 Jennifer R. Wolch, Jason Byrne, and Joshua P. Newell, "Urban green space, public health, and environmental justice: The challenge of making cities 'just green enough,'" *Landscape and Urban Planning* 125 (2014): 234–244.

16 City of Portland Bureau of Environmental Services, *Portland's Green Infrastructure: Quantifying the Health, Energy, and Community Livability Benefits* (Portland, OR: ENTRIX, Inc., 2010), accessed November 2016, https://www.portland oregon.gov/bes/article/298042.

References

Streets as Ecosystems

STREETS ARE ECOSYSTEMS

Foster, Josh, Ashley Lowe, and Steve Winkelman. "The value of green infrastructure for urban climate adaptation." *Center for Clean Air Policy* 750 (2011).

Green, Tom L., Jakub Kronenberg, Erik Andersson, Thomas Elmqvist, and Erik Gómez-Baggethun. "Insurance value of green infrastructure in and around cities." *Ecosystems* (2016): 1–13.

Lukes, Robb, and Christopher Kloss. "Managing wet weather with green infrastructure, Municipal Handbook, Green Streets." *Low Impact Development Center,* Vol. 9. EPA-833-F-08 (2008), Environmental Protection Agency.

Netusil, Noelwah R., Zachary Levin, Vivek Shandas, and Ted Hart. "Valuing green infrastructure in Portland, Oregon." *Landscape and Urban Planning* 124 (2014): 14–21.

"The relevance of street patterns and public space in urban areas." UN-Habitat Working Paper (2013): http://mirror .unhabitat.org/downloads/docs/StreetPatterns.pdf.

Rosenzweig, C., W. Solecki, and New York City Panel on Climate Change. "Climate risk information 2013: Observations, climate change projections, and maps." *New York City Panel on Climate Change* (June 2013). http://www. nyc. gov/html/planyc2030 /downloads/pdf/npcc_climate_risk_ informatio (2013).

Walsh, Christopher J., Derek B. Booth, Matthew J. Burns, Tim D. Fletcher, Rebecca L. Hale, Lan N. Hoang, Grant Livingston et al. "Principles for urban stormwater management to protect stream ecosystems." *Freshwater Science* 35, no. 1 (2016): 398–411.

Cities and Climate Change: An Urgent Agenda. The International Bank for Reconstruction and Development/The World Bank (2010).

Planning For Stormwater

DEVELOPING A SUSTAINABLE STORMWATER NETWORK

Andersson, Erik, Stephan Barthel, Sara Borgström, Johan Colding, Thomas Elmqvist, Carl Folke, and Åsa Gren. "Reconnecting cities to the biosphere: Stewardship of green infrastructure and urban ecosystem services." *Ambio* 43, no. 4 (2014): 445–453.

Dunec, JoAnne L. "Banking on green: A look at how green infrastructure can save municipalities money and provide economic benefits community-wide." (2012): 62–63.

Kjellstrom, Tord, Sharon Friel, Jane Dixon, Carlos Corvalan, Eva Rehfuess, Diarmid Campbell-Lendrum, Fiona Gore, and Jamie Bartram. "Urban environmental health hazards and health equity." *Journal of Urban Health* 84, no. 1 (2007): 86–97.

Lennon, Michael, Scott, Mark and O'Neill, Eoin. "Urban design and adapting to flood risk: therole of green infrastructure." *Journal of Urban Design* 19, no. 5 (2014): 745–758.

Liu, Wen, Weiping Chen, and Chi Peng. "Assessing the effectiveness of green infrastructures on urban flooding reduction: A community scale study." *Ecological Modelling* 291 (2014): 6–14.

Lovell, Sarah Taylor, and John R. Taylor. "Supplying urban ecosystem services through multifunctional green infrastructure in the United States." Landscape ecology 28, no. 8 (2013): 1447–1463.

Pitt, Robert. "Small storm hydrology and why it is important for the design of stormwater control practices." *Advances in modeling the management of stormwater impacts* 7 (1999): 61–91.

Ruckelshaus, Mary H., Gregory Guannel, Katherine Arkema, Gregory Verutes, Robert Griffin, Anne Guerry, Jess Silver, Joe Faries, Jorge Brenner, and Amy Rosenthal. "Evaluating the benefits of green infrastructure for coastal areas: Location, location, location." *Coastal Management* 44, no. 5 (2016): 504–516.

Stormwater Streets

STORMWATER STREET TYPES

Anderson, Sarah, and Holly Piza. *Ultra-Urban Green Infrastructure Guidelines*. The City and County of Denver Public Works (2015).

Boston Complete Streets Design Guide. Boston Transportation Department, Boston, MA: 2013.

Chicago Department of Transportation. *Complete Streets Chicago: Design Guidelines*. City of Chicago, IL: 2013.

Deakin, Elizabeth, and Christopher Porter. "Transportation Impacts of Smart Growth and Comprehensive Planning Initiatives." (2004).

Despins, Chris, Robb Lukes, Kyle Vander Linden, Phil James, Christine Zimmer, and Tyler Babony. *Grey to Green Road Retrofits*. Credit Valley Conservation (2013).

Green Infrastructure and Climate Change: Collaborating to Improve Community Resiliency, US Environmental Protection Agency (2016).

Hair, Lisa, and Melissa Kramer. City Green: Innovative Green Infrastructure Solutions For Downtowns And Infill Locations. US Environmental Protection Agency, Office of Sustainable Communities (2016).

Hendrickson, Kenneth, Dominique Lueckenhoff, Christopher Kloss, and Tamara Mittman. *Conceptual Green Infrastructure Design in the Point Breeze Neighborhood, City of Pittsburgh*. US Environmental Protection Agency, Green Infrastructure Community Partner Program (2015).

Keane, Tim et al. *Move Atlanta: a Design Guide for Active, Balanced & Complete Streets*. City of Atlanta (2016).

MacAdam, James. *Green Infrastructure for Southwestern Neighborhoods*. Watershed Management Group (2012). Accessed Nov 2016: https://wrrc.arizona.edu/sites/wrrc.arizona.edu/files/WMG_Green%20Infrastructure%20for%20South western%20Neighborhoods.pdf.

City of Philadelphia, *Green Streets Design Manual.* Philadelphia, PA: 2014. Accessed Dec. 2016 (http://www.phillywatersheds.org/img/GSDM/GSDM_FINAL_20140211.pdf)

Phillips, Ann Audrey. *City of Tucson Water Harvesting Guidance Manual* (2005).

Stack, Rebecca C., Greg Hoffmann, and Brian Van Wye. *Stormwater Management Guidebook*. District Department of Transportation, Watershed Protection Division (2013).

Strecker, Eric, Aaron Poresky, Robert Roseen, Jane Soule, Venkat Gummadi, Rajesh Dwivedi, Adam Questad et al. *Volume Reduction of Highway Runoff in Urban Areas: Final Report*.NCHRP Report 802, Project 25–41. (2014).

Trice, Amy. "Daylighting streams: breathing life into urban streams and communities." American Rivers, Washington (2013).

Venner, Marie, Marc Leisenring, Eric Strecker, and Dan Pankani. *Current Practice of Post-Construction Structural Stormwater Control Implementation for Highways*. No. NCHRP 25-25 Task 83. (2013).

Stormwater Elements

GREEN STORMWATER ELEMENTS

Jeanjean, A. P. R., P. S. Monks, and R. J. Leigh. "Modelling the effectiveness of urban trees and grass on $PM_{2.5}$ reduction via dispersion and deposition at a city scale." *Atmospheric Environment* 147 (2016): 1–10.

"EPA Cool Pavements Compendium." US Environmental Protection Agency, Washington, DC: accessed September 20, 2015. http://www.epa.gov/heatisld/resources/pdf/CoolPaves Compendium.pdf.

McKeand, Tina and Shirley Vaughn. *Stormwater to Street Trees: Engineering Urban Forests for Stormwater Management*. US Environmental Protection Agency, Office of Wetlands, Oceans and Watersheds (2013).

New York City Department of Environmental Protection, *Guidelines for the Design and Construction of Stormwater Management Systems* (2012). Accessed Nov 2016.

GREEN INFRASTRUCTURE CONFIGURATIONS

Portland Bureau of Environmental Services, *City of Portland Stormwater Management Manual* (2016). Accessed Nov 2016.

Portland Bureau of Environmental Services, *Stormwater Curb Extensions Design Manual* (2013). Accessed Nov 2016.

BIORETENTION DESIGN CONSIDERATIONS

Clapp, J. Casey, H. Dennis P. Ryan III, Richard W. Harper, and David V. Bloniarz. "Rationale for the increased use of conifers as functional green infrastructure: A literature review and synthesis." Arboricultural Journal: The International Journal of Urban Forestry 36, no. 3 (2014): 161–178.

Kondo, Michelle C., Raghav Sharma, Alain F. Plante, Yunwen Yang, and Igor Burstyn. "Elemental concentrations in urban green stormwater infrastructure soils." *Journal of Environmental Quality* 45, no. 1 (2016): 107–118.

Phillips, Tom, Peg Staeheli, Bruce Meyers, Dick Lilly, NancyEllen Regier, and Anthony Harris, *Seattle's Natural Drainage Systems.* Seattle Public Utilities (2013).

Montgomery, James A., Christie A. Klimas, Joseph Arcus, Christian DeKnock, Kathryn Rico, Yarency Rodriguez, Katherine Vollrath, Ellen Webb, and Allison Williams. "Soil quality assessment is a necessary first step for designing urban green infrastructure." *Journal of Environmental Quality* 45, no. 1 (2016): 18–25.

Partnerships & Performance

POLICIES, PROGRAMS, & PARTNERSHIPS

Dunn, A.D., 2010. *Siting green infrastructure: legal and policy solutions to alleviate urban poverty and promote healthy communities.* Boston College Environmental Affairs Law Review, 37.

Green Stormwater Infrastructure in Seattle: Implementation Strategy 2015–2020. City of Seattle (2015).

Green Stormwater Infrastructure Maintenance Manual. City of Austin, Department of Watershed Protection (2016).

Hall, Abby. "Green infrastructure case studies: municipal policies for managing stormwater with Green Infrastructure." Retrieved from United States Environmental Protection Agency: http://rfcd. pima. gov/pdd/lid/pdfs/40-usepa-gi -casestudies-2010. pdf (2010).

Philadelphia Water Department, *Green Stormwater Infrastructure Maintenance Manual.* Philadelphia, PA: 2014.

PERFORMANCE MEASURES

Econsult Solutions, *The Economic Impact of Green City, Clean Waters* (2016).

Gill, Susannah E., John F. Handley, A. Roland Ennos, and Stephan Pauleit. "Adapting cities for climate change: the role of the green infrastructure." *Built Environment* 33, no. 1 (2007): 115–133.

Gómez-Baggethun, Erik, Åsa Gren, David N. Barton, Johannes Langemeyer, Timon McPhearson, Patrick O'Farrell, Erik Andersson, Zoé Hamstead, and Peleg Kremer. "Urban ecosystem services." In *Urbanization, Biodiversity and Ecosystem Services: Challenges and opportunities*, pp. 175–251. Springer Netherlands, 2013.

Hanley, Michael E., and Dave Goulson. "Introduced weeds pollinated by introduced bees: Cause or effect?." *Weed Biology and Management* 3, no. 4 (2003): 204–212.

Houston, Douglas, Jun Wu, Paul Ong, and Arthur Winer. "Structural disparities of urban traffic in southern California: Implications for vehicle-related air pollution exposure in minority and high-poverty neighborhoods." *Journal of Urban Affairs* 26, no. 5 (2004): 565–592.

Jones, Matthew, and John McLaughlin. Green Infrastructure in *New York City: Three Years of Pilot Implementation and Post-Construction Monitoring* (2015).

Landscape Architecture Foundation. "Landscape Performance Series." Accessed Dec. 2016 (http://landscapeperformance .org/).

Lee, Andrew CK, and R. Maheswaran. "The health benefits of urban green spaces: a review of the evidence." *Journal of Public Health* 33, no. 2 (2011): 212–222.

New York City Department of Transportation, *The Economic Benefits of Sustainable Streets.* New York City, NY: 2013.

Vanderwarker, Amy. "Water and Environmental Justice." *A Twenty-First Century US Water Policy* (2012): 52.

City Policy & Guidance References

While not comprehensive, below are examples of city and regional policies pertaining to green stormwater management and multi-modal street design. Refer to page 116 for an in-depth discussion of policy.

	Policy / Plan Type	Example Cities
Policy / Plan	**Complete Streets Policy**	» San Francisco Better Streets Policy » Seattle Complete Streets Policy
	Green Stormwater Infrastructure Plan	» Atlanta Green Infrastructure Action Plan » Philadelphia Green City, Clean Waters
	Green Streets Plan / Policy	» Portland Green Street Policy
Design Guidance / Standards	**Complete Street Design Guidance / Manual**	» Boston Complete Streets Design Guide » New York City Street Design Manual » Philadelphia Complete Streets Design Handbook » San Francisco Better Streets Plan
	Stormwater Management Standards / Manual	» Chattanooga Rainwater Management Guide » Portland Stormwater Management Manual » Seattle Stormwater Manual » San Francisco Stormwater Management Requirements and Design Guidelines » Washington, D.C. Stormwater Management Guidebook
	Green Street Design Guidance	» Chicago Sustainable Urban Infrastructure Guidelines » Denver Ultra-Urban Green Infrastructure Guidelines » Milwaukee Green Streets Stormwater Management Plan » Philadelphia Green Streets Design Manual
	Standard Drawings Set	» Austin » San Francisco » New York City » Vancouver » Philadelphia » Ventura, CA » Washington, D.C.
Stormwater Code / Ordinace	**Stormwater Code / Regulations / Ordinance / Zoning**	» Los Angeles Low Impact Development Ordinance » Minneapolis Stormwater Management Ordinance » Seattle Stormwater Code
Incentives	**Development Fee / Incentive**	» Chicago Green Permit Program » Philadelphia Green Project Review » Seattle Green Factor Program
	Stormwater Fee Discount / Credit	» Detroit Drainage Charge Green Infrastructure Credit program » Minneapolis Stormwater Utility Fee Credit program » Washington, D.C. RiverSmart Rewards incentive program

Credits

Project Steering Committee

ATLANTA, GA

Andrew Walter, RLA, Atlanta Department of Public Works
Cory Rayburn, CPESC, CFM, Atlanta Department of Watershed Management

AUSTIN, TX

Tom Franke, EIT, Austin Watershed Protection Department

BURLINGTON, VT

Megan Moir, Burlington Public Works Department

CAMBRIDGE, MA

Catherine Daly Woodbury, Cambridge Department of Public Works

CHARLOTTE, NC

Johanna Quinn, Charlotte Department of Transportation

CHATTANOOGA, TN

Mark Heinzer, PE, LEED AP, CPESC, CMS4S, City of Chattanooga
Greg Herold, Chattanooga Transportation Department
Cortney Geary, Chattanooga-Hamilton County Transportation Planning Organization

CHICAGO, IL

Hannah Higgins, ASLA, Chicago Department of Transportation
Dave Seglin, Chicago Department of Transportation

DENVER, CO

Sarah Anderson, City and County of Denver Public Works Department
Holly Piza, PE, Urban Drainage and Flood Control District

DETROIT, MI

Janet Attarian, Detroit Planning & Development Department

EL PASO, TX

Fred Lopez, AICP, CNU-A, El Paso Capital Improvement Department
James Fisk, AICP, CNU-A, El Paso Capital Improvement Department
Nicole Ferrini, LEED AP BD+C, CNU-A, El Paso Office of Resilience & Sustainability
Lauren Baldwin, LEED-GA, El Paso Office of Resilience & Sustainability
Jenny Hernandez, El Paso Office of Resilience & Sustainability

FORT LAUDERDALE, FL

Alia Awwad, Fort Lauderdale Transportation & Mobility Department
Elkin Diaz, MBA, PE, PMP, LEED-GA, Fort Lauderdale Public Works Department

HOUSTON, TX

Rod Pinheiro, Houston Department of Public Works and Engineering

INDIANAPOLIS, IN

Rachel Wilson, PE, Indianapolis Department of Public Works

LOS ANGELES, CA

Deborah Deets, RLA, Los Angeles Bureau of Sanitation
Valerie Watson, Los Angeles Department of Transportation

LOUISVILLE, KY

Jordan A. Basham, Louisville Metropolitan Sewer District
Dirk L. Gowin, PE, PLS, PTOE, Louisville Metro Public Works

MINNEAPOLIS, MN

Paul Hudalla, PE, CFM, Minneapolis Public Works Department
Rebecca Hughes, Minneapolis Department of Community Planning and Economic Development
Lacy Shelby, Minneapolis Department of Community Planning and Economic Development

Project Steering Committee *(continued)*

NEW YORK, NY

Erin Cuddihy, NYC Department of Transportation
Danielle DeOrsey, NYC Department of Transportation
Neil Gagliardi, NYC Department of Transportation
Derick Tonning, NYC Department of Environmental Protection

PALO ALTO, CA

Shahla Yazdy, Palo Alto Planning and Community Environment Department, Transportation Division
Brad Eggleston, Palo Alto Department of Public Works

PHILADELPHIA, PA

Ariel Ben-Amos, Philadelphia Water Department
Elizabeth Anne Lutes, EIT, Philadelphia Water Department

PITTSBURGH, PA

Katherine Camp, Pittsburgh Water and Sewer Authority
Joshua Lippert, Pittsburgh Department of City Planning
Megan Zeigler, Pittsburgh Water and Sewer Authority

PORTLAND, OR

Nicole Blanchard, PE, Portland Bureau of Transportation
Ivy Dunlap, RLA, Portland Bureau of Environmental Services
Kate Hibschman, RLA, Portland Bureau of Environmental Services

SAN DIEGO, CA

Eric Mosolgo, San Diego Transportation and Storm Water Department

SAN FRANCISCO, CA

Mike Adamow, San Francisco Public Utilities Commission
John Dennis, PLA, San Francisco Department of Public Works
Robin Welter, RLA, San Francisco Department of Public Works

SALT LAKE CITY, UT

Jason Draper, PE, CFM, Salt Lake City Public Utilities
Lani Kai Eggertsen-Goff, MS, AICP, Salt Lake City Department of Community and Neighborhoods, Engineering Division
Alexis Verson, Salt Lake City Department of Community and Neighborhoods, Transportation Division

SAN JOSE, CA

Ralph Mize, City of San José

SEATTLE, WA

Shanti Colwell, PE, Seattle Public Utilities
Susan McLaughlin, Seattle Department of Transportation

VANCOUVER, WA

Jennifer Campos, Vancouver Community and Economic Development Department
Patrick Sweeney, City of Vancouver

VENTURA, CA

Tom Mericle, PE, TE, Ventura Public Works Department

WASHINGTON, DC

Meredith Upchurch, District Department of Transportation

WEST HOLLYWOOD, CA

Robyn Eason, West Hollywood Community Development Department
Walter Davis, West Hollywood Community Development Department

Contributors

Shanti Colwell, PE, Seattle Public Utilities

Susan McLaughlin, Seattle Department of Transportation

Lacy Shelby, Minneapolis Department of Community
Planning and Economic Development

Ariel Ben-Amos, Philadelphia Water Department

Consultant Team

Peg Staeheli, PLA, FASLA, LEED AP, MIG | SvR

Kathryn Gwilym, PE, LEED AP, MIG | SvR

Amalia Leighton, PE, AICP, MIG | SvR

Nathan Polanski, PE, MIG | SvR

Technical Review Team

Michelle Adams, PE, Meliora Environmental Design

Scott Struck, Ph.D., Geosyntec Consultants

Neil Weinstein, PE, Low Impact Development Center

Jason Wright, PE, Tetra Tech

Partner Organizations

American Society of Civil Engineers

Island Press

Seattle Public Utilities

Summit Foundation

Photo Credits

Atlanta Dept. of Public Works & Dept. of Watershed Management: 90–91 (all)

Ben Baldwin: 97 (boarding island, Portland)

Nicole Blanchard: 97 (bus bulb, SE Division St)

Cambridge Department of Public Works: 107 (Larch St); 108 (splash pad)

CannonCorp Engineering: 13 (21st Street)

Adams Carroll (Flickr user): 41 (28th-31st Ave Connector)

Center for Neighborhood Technology: 109 (Aurora, IL)

Chicago Department of Transportation: 52–53 (all Argyle Street); 65 (Cermak public realm & bike lane pavers); 136 (Greencorps)

Christopher Burke Engineering: 10 (Lawrence St)

Shanti Colwell: 6 (green bump-out); 47 (residential swale); 83 (bioretention swale); 111 (flowers)

Delaware River Waterfront: 73 (Penn Street Trail, before & after)

Jym Dyer: 33 (Oak Street bikeway planter, Fell Street planter, and Fell Street fence)

Kate Fillin-Yeh: 71 (47th & Euclid); 79 (informational sign)

Nathaniel Fink: 89 (Western Ave)

Joe Gilpin: 21 (Tuscon); 111 (xeriscape)

Chris Hamby: 24 (New York)

Greg Herold: 60–61 (all Johnson St)

Infrogmation (Flickr user): 17 (St. Charles Avenue)

Los Angeles Department of Sanitation: 20–21 (all images, with Cal Poly-Pomona); 59 (green alley); 68–69 (all Ed P Reyes River Greenway)

Louisville MSD: 87 (Story Ave)

Tom Mericle: 88 (porous parking lane); 106 (curb cut)

MetroTransit: 26 (University Avenue)

MIG | SvR: 99 (21st Street median); 106 (all Seattle inlets); 109 (Seattle & Kansas City presettlements)

Mike Nakamura: 5 (Vine Street); 74 (Yale Ave N); 80 (seating); 86 (Boren Ave N); 106 (covered inlets); 115 (Yale Ave N)

Nashville MPO: 40–41 (28th-31st Avenue Connector & transit shelter)

New York City Department of Environmental Protection: 79 (stone forebay); 104 (depressed curb); 118–119 (all); 130 (monitoring); 133 (Green Infrastructure map)

New York City Parks & Recreation Department: 132 (NYC tree map)

New York City Department of Transportation: 7 (Allen Street transformation); 97 (planters behind boarding platform)

Philadelphia Water Department: 1 (Fairmount Avenue & N 3rd Street); 72–73 (all Washington Ln & Stenton Ave, Trenton & Norris); 79 (planter step-out); 89 (Percy St); 95 (bump-out); 112 (Schuylkill River Park); 124 (maintenance); 132 (greened acre)

City of Phoenix: 137 (heat island map)

Portland Bureau of Transportation: 103 (SE Tacoma Ave); 107 (SE Tacoma Ave & concrete apron); 130 (SW Montgomery St)

Portland Metro: 45 (all); 85 (SE Division St); 137 (green curb extension)

Dan Reed: 93 (Washington, DC); 106 (trenched inlet)

BL Ross: 89 (Indy Cultural Trail)

Sergio Ruiz: 32 (Oak Street, primary photo)

Julianne Sabula: 17 (Sugar House Streetcar)

San Francisco Public Utilities Commission: 33 (Fell St planter & crosswalk); 77 (Newcomb Ave); 134 (Newcomb Ave)

Seattle Public Utilities: 48–49 (all); 56–57 (all); 83 (swale); 111 (tall grasses)

Lacy Shelby: 83 (grass swale); 109 (Austin presettlement structure)

Photo Credits *(continued)*

TriMet: 29 (SW Lincoln Ave)

Edward Tuene (via Wikimedia Commons): 131 (butterfly on milkweed)

Steven Vance: 64 (Cermak St), 65 (solar/wind collection)

Aaron Villere: 44 (SE Division St), 95 (SE Division St); 107 (Tilikum Crossing); 109 (Portland); 111 (sedum)

Cherie Walkowiak, Safe Mountain View: 98 (Rosemead Blvd); 95 (Palo Alto)

Eric Wheeler: 17 (University Avenue); 25 (University Avenue, St. Paul); 33–37 (all); 127 (Washington Avenue)